Springer Optimization and Its Applications

Volume 192

Aims and Scope

Optimization has continued to expand in all directions at an astonishing rate. New algorithmic and theoretical techniques are continually developing and the diffusion into other disciplines is proceeding at a rapid pace, with a spot light on machine learning, artificial intelligence, and quantum computing. Our knowledge of all aspects of the field has grown even more profound. At the same time, one of the most striking trends in optimization is the constantly increasing emphasis on the interdisciplinary nature of the field. Optimization has been a basic tool in areas not limited to applied mathematics, engineering, medicine, economics, computer science, operations research, and other sciences.

The series **Springer Optimization and Its Applications (SOIA)** aims to publish state-of-the-art expository works (monographs, contributed volumes, textbooks, handbooks) that focus on theory, methods, and applications of optimization. Topics covered include, but are not limited to, nonlinear optimization, combinatorial optimization, continuous optimization, stochastic optimization, Bayesian optimization, optimal control, discrete optimization, multi-objective optimization, and more. New to the series portfolio include Works at the intersection of optimization and machine learning, artificial intelligence, and quantum computing.

Volumes from this series are indexed by Web of Science, zbMATH, Mathematical Reviews, and SCOPUS.

Yuri Sachkov

Introduction to
Geometric Control

 Springer

Yuri Sachkov
Program Systems Institute
Pereslavl-Zalessky
Russia

English translation of Geometric Control Theory, published in Russian, Moscow, URSS, 2021

ISSN 1931-6828   ISSN 1931-6836   (electronic)
Springer Optimization and Its Applications
ISBN 978-3-031-02069-8   ISBN 978-3-031-02070-4   (eBook)
https://doi.org/10.1007/978-3-031-02070-4

Mathematics Subject Classification: 93, 49, 53C17

This Springer imprint is published by the registered company Springer Nature Switzerland AG
The registered company address is: Gewerbestrasse 11, 6330 Cham, Switzerland

To Elena

Preface

Alone in the wilderness, lost in the jungle, the boy is searching, searching!
The swelling waters, the far-away mountains, and the unending path;
Exhausted and in despair, he knows not where to go,
He only hears the evening cicadas singing in the maple-woods.

Pu-ming, "The Ten Oxherding Pictures" (cited by Suzuki [110])

Searching for the Ox

What Is This Book About?

This is a short introductory course on geometric control theory with emphasis on controllability and optimal control problems. This direction of mathematical control theory originated in the 1970s; it actively employs methods of differential geometry, Lie groups and Lie algebras, and symplectic geometry for the study of control systems. The main literature on this subject are the books by V. Jurdjevic [26], A. Agrachev and the author [3], H. Schättler and U. Ledzewicz [34], and a very recent book by A. Agrachev, D. Barilari, and U. Boscain [2] (in chronological order).

The course aims to give a very concise first impression on the circle of problems, methods, and results of geometric control theory, and to advance problem solving skills employing these methods and results.

Whom Is This Book for?

The book is intended for a wide readership, starting from undergraduate students in mathematics and applied mathematics to experts in "classic" control theory, for whom geometric methods might be new.

What Prerequisites Are Assumed?

The working knowledge of the basic courses taught in undergraduate mathematics programmes, such as calculus, linear algebra, and ordinary differential equations, is required.

Acquaintance with smooth manifolds and vector fields on them, and elements of Lie theory is desirable (but not necessary).

Knowledge of control theory is not expected.

What Is the Purpose of This Book?

The book may serve as a textbook for a semester or half-semester special course for students starting from the undergraduate level, as was done by the author in universities of Pereslavl-Zalessky, Moscow, Novosibirsk, Sochi, Trieste, Rouen, Brno, Braşov, Jyväskylä, Changchun, and Shanghai.

Pedagogical Principles

We aim to achieve the following goals in this book:

- Low initial prerequisites for readers
- Problem-oriented exposition
- Theoretical minimalism: only the theory absolutely necessary to start geometric control is included
- Active use of illustrations
- Relatively concise volume
- Inclusion of some advanced, recently published material

Structure of the Book

The book consists of four chapters, Conclusion, and an appendix.

Chapter 1 is introductory and is devoted to acquainting the reader with control problems and smooth manifolds. In Sect. 1.1, we state several particular geometric control problems: stopping a train, the Markov–Dubins car, the sub-Riemannian problem on the group of motions of the plane, Euler's elasticae, the plate-ball problem, anthropomorphic curve restoration, and Dido's problem. Next, we describe the main general problems of the course—the controllability problem and the optimal control problem. In Sect. 1.2, we recall initial facts about smooth manifolds, vector fields, and Lie groups required in the book.

Chapter 2 is focused on the controllability problem. In Sect. 2.1, we prove some classic results on this problem: Kalman's controllability test for linear systems and the sufficient local controllability condition of a nonlinear system via linear approximation. In Sect. 2.2, we prove the fundamental Nagano–Sussmann Orbit theorem and its important corollaries: the Rashevskii–Chow theorem and the Frobenius theorem. Section 2.3 is devoted to the famous Krener's theorem on the structure of attainable sets of full-rank systems.

Chapter 3 is devoted to the optimal control problem. In Sect. 3.1, we study this second main problem of the course: we state the problem and discuss Filippov's theorem—a fundamental sufficient condition for existence of optimal controls. Section 3.2 is dedicated to the main necessary optimality condition in optimal control problems—the Pontryagin maximum principle. We use the symplectic geometry formalism to state the coordinate-free version of this theorem for problems on smooth manifolds. In Sect. 3.3, we consider sub-Riemannian problems, a popular domain during last decades. In this section, preceding results are specialized for these problems, and optimality conditions are discussed (conjugate points, Maxwell points). Additionally, the Pontryagin maximum principle for sub-Riemannian problem is proved. In Sect. 3.4, we describe a general symmetry method for constructing optimal synthesis in optimal control problems with a big symmetry group.

In Chap. 4, we present solutions to several important left-invariant optimal control problems on Lie groups: Dido's problem, Euler's elastic problem, and the sub-Riemannian problems on the group of motions of the plane and on the Engel group.

The Conclusion contains recommendations for further reading.

In the appendix, we collect some basic facts on elliptic functions and integrals, and on the equation of pendulum.

In the majority of sections, theoretical results are illustrated by the examples of problems stated in Sect. 1.1.1. Some sections include exercises.

The epigraphs to the preface and chapters of this book are poems by Pu-min that accompany "Ten Oxherding Pictures" of Zen Buddhism [110].

Acknowledgments The author is grateful to his teacher Andrei Alexandrovich Agrachev for continuous lessons of mathematics, control theory, and life.

The author thanks his students Alexey Mashtakov, Alexey Podobryaev, and Andrei Ardentov for their careful reading of the text and a series of valuable suggestions.

The author is grateful to Elena Sachkova for the help with drawing figures.

Also, the author wishes to thank Ms Elizabeth Loew (Springer) for extensive editorial help with publication of this book.

Pereslavl-Zalessky, Russia   Yuri Sachkov
March 2022

Contents

List of Figures

Chapter 1
Introduction

By the stream and under the trees, scattered are the traces of the lost;
The sweet-scented grasses are growing thick—did he find the way?
However remote over the hills and far away the beast may wander,
His nose reaches the heavens and none can conceal it.

Pu-ming, "The Ten Oxherding Pictures" (cited by Suzuki [110])

Seeing the Traces

We begin this chapter by stating several specific control problems, many of which will be studied in the book. Then we describe two main general problems of this book—the controllability problem and the optimal control problem. Finally, we recall some basic facts on smooth manifolds, vector fields, and Lie groups.

© The Author(s), under exclusive license to Springer Nature Switzerland AG 2022
Yu. Sachkov, *Introduction to Geometric Control*, Springer Optimization
and Its Applications 192, https://doi.org/10.1007/978-3-031-02070-4_1

1.1   Statement of Control Problems

1.1.1   Examples of Optimal Control Problems

Before developing the general theory, we state several particular optimal control problems.

1.1.1.1   Stopping a Train

Consider a material point of mass $m > 0$ with coordinate $x \in \mathbb{R}$ that moves along a line under the action of a force F bounded by the absolute value by $F_{\max} > 0$. Given an initial position x_0 and initial velocity $\dot{x}_0$ of the material point, we should find a force F that steers the point to the origin with zero velocity, for a minimal time.

The second law of Newton gives $|m\ddot{x}| = |F| \le F_{\max}$, thus $|\ddot{x}| \le \frac{F_{\max}}{m}$. Choosing appropriate units of measure, we can obtain $\frac{F_{\max}}{m} = 1$, thus $|\ddot{x}| \le 1$. Denote position of the point $x_1 = x$, velocity $x_2 = \dot{x}$, and acceleration $u = \ddot{x}$, $|u| \le 1$. Then the problem is formalized as follows:

$$\dot{x}_1 = x_2, \qquad (x_1, x_2) \in \mathbb{R}^2, \tag{1.1}$$

$$\dot{x}_2 = u, \qquad |u| \le 1, \tag{1.2}$$

$$(x_1, x_2)(0) = (x_0, \dot{x}_0), \qquad (x_1, x_2)(t_1) = (0, 0),$$

$$t_1 \to \min.$$

Such an optimal control problem, where one should minimize the time t_1 of motion of the system between two given points, is called a *time-optimal problem*. The problem is linear since the right-hand side of equations (1.1), (1.2) depends linearly both on the *state* (x_1, x_2) and the *control parameter* u. This problem is studied in Sect. 3.2.3.1.

1.1.1.2   Control of Linear Oscillator

Consider a pendulum that performs small oscillations under the action of a force bounded by the absolute value. We should choose a force that steers the pendulum from an arbitrary position and velocity to the stable equilibrium for a minimum time. After choosing appropriate units of measure, we get a mathematical model:

$$\ddot{x}_1 = -x_1 + u, \qquad |u| \le 1, \quad x_1 \in \mathbb{R}.$$

Introducing the notation $x_2 = \dot{x}_1$ for velocity, we get a linear time-optimal problem:

$$\dot{x}_1 = x_2, \qquad x = (x_1, x_2) \in \mathbb{R}^2,$$
$$\dot{x}_2 = -x_1 + u, \qquad |u| \le 1,$$
$$x(0) = x^0, \qquad x(t_1) = 0,$$
$$t_1 \to \min.$$

This problem is discussed in Sect. 3.2.3.3.

1.1.1.3   The Markov–Dubins Car

Consider a simplified model of a car given by a unit vector attached at a point $(x, y) \in \mathbb{R}^2$, with orientation $\theta \in S^1$ (here and below $S^1 = \mathbb{R}/(2\pi\mathbb{Z})$ is the circle, i.e., the one-dimensional sphere). The car moves forward with the unit velocity and can simultaneously rotate with an angular velocity $|\dot{\theta}| \le 1$. Given an initial and a terminal state of the car, we should choose the angular velocity in such a way that the time of motion is as minimum as possible.

We have the following time-optimal problem:

$$\dot{x} = \cos\theta, \qquad q = (x, y, \theta) \in \mathbb{R}^2_{x,y} \times S^1_\theta = M,$$
$$\dot{y} = \sin\theta, \qquad |u| \le 1,$$
$$\dot{\theta} = u,$$
$$q(0) = q_0, \qquad q(t_1) = q_1,$$
$$t_1 \to \min.$$

This problem is nonlinear, moreover, the *state space* $M = \mathbb{R}^2 \times S^1$ is a nontrivial smooth manifold, homeomorphic to the solid torus. Parallel translations and rotations (i.e., motions of the plane) transform solutions of the problem into solutions, thus the problem is *left-invariant* on the *group of motions of Euclidean plane* SE(2) $\cong \mathbb{R}^2 \times S^1$; see details on this group in Sect. 4.2. The problem on the Markov–Dubins car is studied in Sect. 3.2.3.2.

1.1.1.4   The Sub-Riemannian Problem on the Group of Motions of the Plane

Consider a (more realistic) model of a car in the plane that can move forward or backward with an arbitrary linear velocity and simultaneously rotate with an arbitrary angular velocity. The state of the car is given by its position in the plane and orientation angle. We should find a motion of the car from a given initial state to a given terminal state, so that the length of the path in the space of positions and orientations is as minimum as possible; see Fig. 1.1.

We get the following optimal control problem:

$$\dot{x} = u \cos \theta, \qquad q = (x, y, \theta) \in \mathbb{R}^2_{x,y} \times S^1_\theta, \tag{1.3}$$

$$\dot{y} = u \sin \theta, \qquad (u, v) \in \mathbb{R}^2, \tag{1.4}$$

$$\dot{\theta} = v, \tag{1.5}$$

$$q(0) = q_0, \qquad q(t_1) = q_1,$$

$$l = \int_0^{t_1} \sqrt{\dot{x}^2 + \dot{y}^2 + \dot{\theta}^2} \, dt = \int_0^{t_1} \sqrt{u^2 + v^2} \, dt \to \min. \tag{1.6}$$

This is a nonlinear optimal control problem with integral *cost functional l*. Dynamics (1.3)–(1.5) is linear in the control (u, v), while the cost functional l is homogeneous of order 1 in the control, so we have a *sub-Riemannian problem*; see Sect. 3.3. Motions of the plane transform solutions into solutions, thus the problem is left-invariant on the Lie group SE(2). This problem is considered in detail in Sect. 4.2.

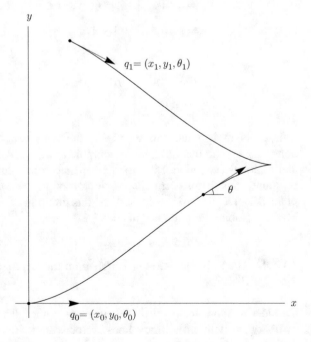

Fig. 1.1 Sub-Riemannian problem on the group of motions of the plane

1.1.1.5   Euler Elasticae

Consider a uniform elastic rod of length l in the plane. Suppose that the rod has fixed endpoints and tangents at endpoints. We should find the profile of the rod.

Let $(x(t), y(t))$ be an arclength parametrization of the rod, and let $\theta(t)$ be its orientation angle in the plane. Then the form of the rod $\gamma(t) = (x(t), y(t))$ is determined by the system

$$
\begin{aligned}
\dot{x} &= \cos\theta, & q &= (x, y, \theta) \in \mathbb{R}^2 \times S^1, \\
\dot{y} &= \sin\theta, & u &\in \mathbb{R}, \\
\dot{\theta} &= u, & & \\
q(0) &= q_0, & q(t_1) &= q_1, \quad t_1 = l \text{ is the length of the rod;}
\end{aligned}
$$

see Fig. 1.2, where $q_i = (a_i, \theta_i)$, $i = 0, 1$.

The elastic energy of the rod is equal to $J = \frac{1}{2}\int_0^{t_1} k^2 \, dt$, where k is the curvature of the rod. Since for an arclength parametrised rod $k = \dot{\theta} = u$, we obtain the cost functional

$$
J = \frac{1}{2}\int_0^{t_1} u^2 \, dt \to \min,
$$

because the rod takes the form that minimizes its elastic energy.

Once more we have a nonlinear optimal control problem with integral cost functional, left-invariant on the Lie group SE(2). Although, now the system is affine (linear nonhomogeneous) in the control, thus the problem is not sub-Riemannian. This problem is studied in Sect. 4.3.

1.1.1.6   The Plate-Ball Problem

Let a uniform sphere roll without slipping or twisting on a horizontal plane. One can imagine that the sphere rolls between two horizontal planes: a fixed lower one

Fig. 1.2 Euler's elastic problem

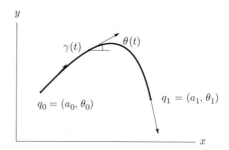

and a moving upper one. Admissible motions are obtained by horizontal motions of the upper plane. Absence of slipping means that the contact point of the sphere with the plane has zero instantaneous velocity; and absence of twisting means that the angular velocity vector of the sphere is horizontal. We should roll the sphere from a given initial state to a given terminal state, so that the length of the curve in the plane traced by the contact point was the minimum possible. We are interested in the kinematics of motion of the sphere, thus we can forget about the upper plane.

State of the system is determined by the contact point of the sphere and the plane, and the orientation of the sphere in the space. Introduce a fixed orthonormal frame (e_1, e_2, e_3) in the space such that e_1 and e_2 are contained in the horizontal plane, and the vector e_3 looks to the half-space where the sphere rolls. Further, let (f_1, f_2, f_3) be a moving orthonormal frame attached to the sphere. Let a point of the sphere have coordinates (x, y, z) in the fixed frame (e_1, e_2, e_3), and coordinates (X, Y, Z) in the moving frame (f_1, f_2, f_3):

$$xe_1 + ye_2 + ze_3 = Xf_1 + Yf_2 + Zf_3.$$

Let (x, y) denote coordinates of the contact point of the sphere with the plane. Then the orthogonal 3×3 matrix R such that

$$R \begin{pmatrix} x \\ y \\ z \end{pmatrix} = \begin{pmatrix} X \\ Y \\ Z \end{pmatrix}$$

determines orientation of the sphere in the space. This matrix belongs to the *group of rotations of the 3-dimensional space*:

$$R \in \mathrm{SO}(3) = \left\{ A \in \mathbb{R}^{3 \times 3} \mid A^T = A^{-1}, \quad \det A = 1 \right\},$$

where $\mathbb{R}^{3 \times 3}$ is the space of all 3×3 real matrices. Then the state of the system is

$$q = (x, y, R) \in \mathbb{R}^2 \times \mathrm{SO}(3) = M,$$

and our problem is written as follows:

$$\dot{x} = u, \quad \dot{y} = v, \qquad (u, v) \in \mathbb{R}^2, \tag{1.7}$$

$$\dot{R} = R \begin{pmatrix} 0 & 0 & -u \\ 0 & 0 & -v \\ u & v & 0 \end{pmatrix}, \tag{1.8}$$

Fig. 1.3 The plate-ball problem

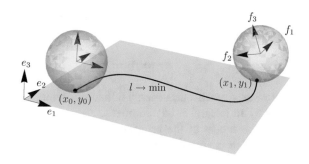

$$q(0) = q_0, \qquad q(t_1) = q_1,$$

$$l = \int_0^{t_1} \sqrt{u^2 + v^2}\, dt \to \min;$$

see Fig. 1.3.

This is a left-invariant sub-Riemannian problem on the 5-dimensional Lie group $\mathbb{R}^2 \times SO(3)$. It was studied in detail in [80, 90]; see also Exercise 8 in Sect. 4.5.

1.1.1.7 Anthropomorphic Curve Reconstruction

Suppose that a greyscale image is given by a set of isophotes (level lines of brightness). Let the image be corrupted in some domain, and our goal is to reconstruct it anthropomorphically, i.e., close to the way a human brain does. Consider a problem of anthropomorphic reconstruction of a curve, i.e., of reconstruction by a method similar to the human one, when we reconstruct a corrupted curve to a complete arc.

According to a discovery of D. Hubel and T. Wiesel [102] (1981 Nobel Prize in Physiology or Medicine), a human brain stores curves not as sequences of planar points (x_i, y_i), but as sequences of positions and orientations (x_i, y_i, θ_i). Moreover, an established model of the primary visual cortex $V1$ of the human brain by J. Petitot, G. Citti and A. Sarti [100, 107] states that corrupted curves of images are reconstructed according to a variational principle, i.e., in a way that minimizes the activation energy of neurons required for drawing the missing part of the curve.

So the discovery by Hubel and Wiesel states that the human brain lifts images $(x(t), y(t))$ from the plane to the space of positions and orientations $(x(t), y(t), \theta(t))$. The lifted curve is a solution to the control system

$$\dot{x} = u\cos\theta, \qquad q = (x, y, \theta) \in \mathbb{R}^2 \times S^1,$$

$$\dot{y} = u\sin\theta, \qquad (u, v) \in \mathbb{R}^2,$$

$$\dot{\theta} = v,$$

with the boundary conditions provided by endpoints and tangents of the corrupted curve:

$$q(0) = q_0, \qquad q(t_1) = q_1.$$

Moreover, the activation energy of neurons required to draw the corrupted curve is given by an integral to be minimized.

By the model of the primary visual cortex V1 due to J. Petitot, G. Citti and A. Sarti, this energy is measured by the integral

$$J = \int_0^{t_1} (u^2 + v^2)\, dt \to \min.$$

By the Cauchy–Schwarz inequality, minimization of the energy J is equivalent to minimization of the length functional

$$l = \int_0^{t_1} \sqrt{u^2 + v^2}\, dt \to \min,$$

for fixed t_1. We have a remarkable fact: optimal trajectories for the sub-Riemannian problem on the group of motions of the plane solve the problem of anthropomorphic curve reconstruction!

1.1.1.8 Dido's Problem

> There bought a space of ground, which (Byrsa call'd,
> From the bull's hide) they first inclos'd, and wall'd.
>
> *Virgil "Aeneid"*

Dido's problem (the *left-invariant sub-Riemannian problem on the Heisenberg group*) is the simplest nontrivial sub-Riemannian problem. It is discussed in almost any textbook or survey on sub-Riemannian geometry or geometric control theory, see e.g. [2, 26, 28, 37]. We do not venture to break this tradition.

Consider the following formalization of an ancient optimization problem going back to nineteenth century BC [28, 111]. Given points $a_0, a_1 \in \mathbb{R}^2$, a Lipschitzian curve $\overline{\gamma} \subset \mathbb{R}^2$ connecting a_1 with a_0, and a number $S \in \mathbb{R}$, one should find the shortest Lipschitzian curve $\gamma \subset \mathbb{R}^2$ connecting a_0 with a_1 for which the closed curve $\gamma \cup \overline{\gamma}$ bounds a domain in $\mathbb{R}^2$ of the algebraic area S; see Fig. 1.4.

Introduce Cartesian coordinates x, y in the plane $\mathbb{R}^2$ with the origin a_0. Then $a_0 = (0, 0)$, $a_1 = (x_1, y_1)$, and the curves γ, $\overline{\gamma}$ are parametrised as

$$\gamma(t) = (x(t), y(t)), \qquad t \in [0, t_1],$$
$$\overline{\gamma}(t) = (\overline{x}(t), \overline{y}(t)), \qquad t \in [0, \overline{t}_1].$$

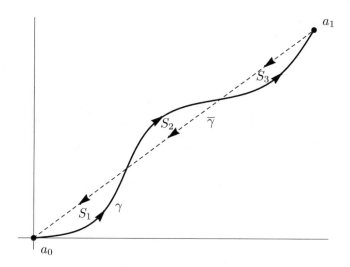

Fig. 1.4 Dido's problem: $S = S_1 - S_2 + S_3 - \dots$, $l(\gamma) \to \min$

The parameters $t_1, \bar{t}_1$ are arbitrary, they can be considered fixed or free since the length of a curve does not depend on parametrisation.

Consider a closed curve $\widehat{\gamma} = \gamma \cup \bar{\gamma}$ and a domain bounded by it: $D \subset \mathbb{R}^2$, $\partial D = \widehat{\gamma}$. Then

$$S(D) = \frac{1}{2} \oint_{\widehat{\gamma}} x\,dy - y\,dx = \frac{1}{2} \int_0^{t_1} (x\dot{y} - y\dot{x})\,dt - \bar{I},$$

$$\bar{I} = \frac{1}{2} \int_0^{\bar{t}_1} (\bar{x}\dot{\bar{y}} - \bar{y}\dot{\bar{x}})\,dt = \frac{1}{2} \int_{\bar{\gamma}} x\,dy - y\,dx.$$

Introduce the variable

$$z(t) := \frac{1}{2} \int_0^t (x\dot{y} - y\dot{x})\,dt$$

that measures the algebraic sectorial area swept by the radius-vector $(x(t), y(t))$. Thus a curve $(x(t), y(t))$ lifts to the curve $(x(t), y(t), z(t))$, where $z(t)$ is equal to the sectorial area $S(t)$; see Fig. 1.5.

Then $z(0) = 0$, and the number $z(t_1) = S + \bar{I} =: z_1$ is given.

Denote functions from the space L^∞:

$$\dot{x}(t) =: u_1(t), \qquad \dot{y}(t) =: u_2(t),$$

Fig. 1.5 Lift of curve
$(x, y)(t)$ to $(x, y, z)(t)$

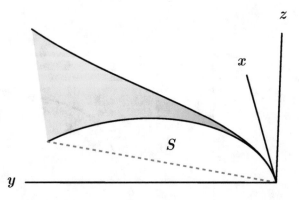

then

$$\dot{z}(t) = \frac{1}{2}(xu_2 - yu_1).$$

Thus along solutions of Dido's problem, the point $q = (x, y, z) \in \mathbb{R}^3$ satisfies the control system

$$\dot{q} = u_1 X_1(q) + u_2 X_2(q), \qquad u = (u_1, u_2) \in \mathbb{R}^2, \quad q \in \mathbb{R}^3, \qquad (1.9)$$

$$X_1 = \frac{\partial}{\partial x} - \frac{y}{2}\frac{\partial}{\partial z}, \qquad X_2 = \frac{\partial}{\partial y} + \frac{x}{2}\frac{\partial}{\partial z}, \qquad (1.10)$$

the boundary conditions

$$q(0) = q_0 = (0, 0, 0), \qquad q(t_1) = q_1 = (x_1, y_1, z_1), \qquad (1.11)$$

and the optimality condition

$$l(\gamma) = \int_0^{t_1} \sqrt{u_1^2 + u_2^2}\, dt \to \min. \qquad (1.12)$$

This is a sub-Riemannian problem on $\mathbb{R}^3$; see Sect. 4.1 for its solution.

By virtue of Exercise 4 (see Sect. 1.3), the length minimization problem (1.12) is equivalent to the energy minimization problem

$$\int_0^{t_1} (u_1^2 + u_2^2)\, dt \to \min.$$

1.1.2 Control Systems and Problems

1.1.2.1 Dynamical Systems and Control Systems

A smooth *dynamical system*, or an *ordinary differential equation (ODE)* on a smooth manifold, is given by an equation

$$\dot{q} = f(q), \qquad q \in M, \tag{1.13}$$

where $f \in \mathrm{Vec}(M)$ is a smooth vector field on M. A basic property of a dynamical system is that it is deterministic, i.e., given an initial condition $q(0) = q_0$ and a time $t > 0$, there exists a unique solution $q(t)$ to ODE (1.13). A typical example of a dynamical system is a planet rotating around the sun.

A *control system* is obtained from dynamical system (1.13) if we add a *control parameter u* in the right-hand side:

$$\dot{q} = f(q, u), \qquad q \in M, \quad u \in U. \tag{1.14}$$

The control parameter varies in a *set of control parameters U* (usually a subset of $\mathbb{R}^m$). This parameter can change in time: we can choose a *control* function $u = u(t) \in U$ and substitute it to the right-hand side of control system (1.14) to obtain a nonautonomous ODE

$$\dot{q} = f(q, u(t)). \tag{1.15}$$

Together with an initial condition

$$q(0) = q_0, \tag{1.16}$$

ODE (1.15) determines a unique solution—a *trajectory* $q_u(t)$, $t \geq 0$, of control system (1.14) corresponding to the control $u(t)$ and initial condition (1.16). For another control $\tilde{u}(t)$, we get another trajectory $q_{\tilde{u}}(t)$ with initial condition (1.16). The manifold M is called the *state space* of control system (1.14). The mapping

$$f : (q, u) \mapsto f(q, u)$$

in (1.14) is assumed smooth.

A typical example of a control system is a rocket flying in the space.

Regularity assumptions for control $u(\cdot)$ can vary from problem to problem; typical examples are piecewise constant controls or Lebesgue measurable bounded controls (from L^∞). The controls considered in a particular problem are called *admissible controls*.

If we fix initial condition (1.16) and vary admissible controls, we get a new object—the *attainable set* of control system (1.14) from the point q_0 for arbitrary times:

$$\mathcal{A}_{q_0} = \{q_u(t) \mid q_u(0) = q_0, \quad u \in L^{\infty}([0, t], U), \quad t \geq 0\}.$$

Notice that in control theory the time parameter t is usually assumed nonnegative. For a dynamical system, the attainable set is not considered since it is just a positive-time half-trajectory. But for control systems, the attainable set is a non-trivial object, and its study is one of the central problems of control theory.

If we apply restrictions on the terminal time of trajectories, we get restricted attainable sets: the attainable set from the point q_0 for a time $t_1 \geq 0$

$$\mathcal{A}_{q_0}(t_1) = \{q_u(t_1) \mid q_u(0) = q_0, \quad u \in L^{\infty}([0, t_1], U)\},$$

and the attainable set from the point q_0 for times not greater than $t_1 \geq 0$:

$$\mathcal{A}_{q_0}(\leq t_1) = \bigcup_{t=0}^{t_1} \mathcal{A}_{q_0}(t).$$

Now we state our two main general problems.

1.1.2.2 The Controllability Problem

Definition 1.1 A control system (1.14) is called:

- *globally (completely) controllable* if $\mathcal{A}_{q_0} = M$ for any $q_0 \in M$
- *globally controllable from a point* $q_0 \in M$ if $\mathcal{A}_{q_0} = M$
- *locally controllable at* q_0 if $q_0 \in \mathrm{int}\,\mathcal{A}_{q_0}$
- *small time locally controllable (STLC) at* q_0 if $q_0 \in \mathrm{int}\,\mathcal{A}_{q_0}(\leq t_1)$ for any $t_1 > 0$,

where int denotes the interior of a set.

Even the local controllability problem is rather hard to solve: there exist necessary conditions and sufficient conditions of STLC for arbitrary dimension of the state space M, but local controllability tests are available only for the case $\dim M = 2$. The global controllability problem is naturally much more harder: there exist global controllability conditions only for very symmetric systems: linear systems, left-invariant systems on Lie groups.

1.1.2.3 The Optimal Control Problem

Suppose that for control system (1.14) the controllability problem between points $q_0, q_1 \in M$ is solved positively: $q_1 \in \mathcal{A}_{q_0}$. Then typically the points q_0, q_1 are connected by more than one trajectory of the control system (usually by continuum of trajectories). Thus there naturally arises the question of the best (optimal in a certain sense) trajectory connecting q_0 and q_1. In order to measure the quality of trajectories (controls), introduce a *cost functional* to be minimized:

$$J = \int_0^{t_1} \varphi(q, u)\, dt.$$

A typical example: we should transfer the space satellite from one position to another with minimal expenditure of fuel.

Thus we get an *optimal control problem*:

$$\dot{q} = f(q, u), \qquad q \in M, \quad u \in U, \tag{1.17}$$

$$q(0) = q_0, \qquad q(t_1) = q_1, \tag{1.18}$$

$$J = \int_0^{t_1} \varphi(q, u)\, dt \to \min. \tag{1.19}$$

Here the *terminal time* t_1 may be fixed or free.

The optimal control problem is also rather hard to solve—this is an optimization problem (1.19) in an infinite-dimensional space $\{u(\cdot)\}$ of admissible controls with specific constraints (1.17), (1.18). There exist general necessary optimality conditions (the most important of which are first order optimality conditions given by the Pontryagin maximum principle) and general sufficient optimality conditions (second-order and higher-order). But optimality tests are available only for special classes of problems (linear, linear-quadratic, convex problems).

So we stated two main general problems of this course:

1. the controllability problem
2. the optimal control problem.

There are many other important mathematical control problems: equivalence, stabilization, observability, etc., which we do not touch upon.

We study the controllability problem in Chap. 2, and the optimal control problem in Chap. 3. The general theory developed is applied to several particular optimal control problems in Chap. 4.

1.2   Smooth Manifolds, Vector Fields, and Lie Groups

Here we recall briefly some basics of calculus on smooth manifolds, vector fields, and Lie groups. For a systematic study we recommend a regular textbook on differential geometry (e.g., [109, 114]).

1.2.1   Smooth Manifolds

A smooth k-dimensional *submanifold* $M \subset \mathbb{R}^n$ is usually defined by one of the following equivalences:

(a) implicitly by a system of regular equations:

$$f_1(x) = \cdots = f_{n-k}(x) = 0, \qquad x \in \mathbb{R}^n,$$

$$\text{rank}\left(\frac{\partial f_1}{\partial x}, \ldots, \frac{\partial f_{n-k}}{\partial x}\right) = n - k,$$

(b) or by a regular parametrization:

$$x = \Phi(y), \qquad y \in \mathbb{R}^k, \quad x \in \mathbb{R}^n,$$

$$\text{rank}\,\frac{\partial \Phi}{\partial y} = k.$$

An abstract smooth *manifold* M (not embedded into $\mathbb{R}^n$) is defined via a system of charts (local coordinates) that mutually agree.

A mapping between smooth manifolds is called *smooth* if it is smooth (of class C^∞) in local coordinates.

The *tangent space* to a smooth submanifold $M \subset \mathbb{R}^n$ at a point $x \in M$ is defined as follows for the two definitions above of a submanifold:

(a) $T_x M = \text{Ker}\,\frac{\partial f}{\partial x}(x)$,
(b) $T_x M = \text{Im}\,\frac{\partial \Phi}{\partial y}(y), x = \Phi(y)$.

Now let M be an abstract smooth manifold and let $q \in M$. Consider a smooth curve $\gamma : (-\varepsilon, \varepsilon) \to M$ with $\gamma(0) = q$. Then the *tangent vector* $\dot\gamma(0) = \frac{d\gamma}{dt}(0)$ is defined as the equivalence class of all smooth curves with $\gamma(0) = q$ and with the same 1-st order Taylor polynomial in some (thus in any) system of local coordinates; see Fig. 1.6.

The *tangent space* to M at a point $q \in M$ is the set of all tangent vectors to M at q:

$$T_q M = \{\dot\gamma(0) \mid \gamma : (-\varepsilon, \varepsilon) \to M \text{ smooth}, \quad \gamma(0) = q\};$$

Fig. 1.6 Tangent vector $\dot\gamma(0)$

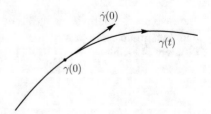

Fig. 1.7 Tangent space T_qM

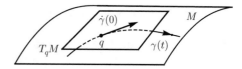

see Fig. 1.7. It is a vector space of the same dimension as M.

Given a smooth mapping $F : M \to N$ between smooth manifolds, for any $q \in M$ the *differential*

$$F_{*q} : T_qM \to T_{F(q)}N$$

is defined as follows:

$$F_{*q}v = \left.\frac{d}{dt}\right|_{t=0} F(\gamma(t)),$$

where $\gamma : (-\varepsilon, \varepsilon) \to M$ is a smooth curve such that $\gamma(0) = q, \dot{\gamma}(0) = v$.

1.2.2   Smooth Vector Fields and Lie Brackets

A smooth *vector field* on a manifold M is a smooth mapping

$$M \ni q \mapsto V(q) \in T_qM.$$

Notation: $V \in \mathrm{Vec}(M)$.

A *trajectory of a vector field* V through a point $q_0 \in M$ is a solution to the Cauchy problem:

$$\dot{q}(t) = V(q(t)), \quad q(0) = q_0.$$

Suppose that a trajectory $q(t)$ exists for all times $t \in \mathbb{R}$, then we denote $e^{tV}(q_0) := q(t)$. The one-parameter group of diffeomorphisms $e^{tV} : M \to M$ is the *flow* of the vector field V.

Now we define the Lie bracket (commutator) of vector fields $V, W \in \mathrm{Vec}(M)$. We say that V and W *commute* if their flows commute:

$$e^{tV} \circ e^{sW} = e^{sW} \circ e^{tV}, \qquad t, s \in \mathbb{R}.$$

In the general case vector fields V and W do not commute, thus $e^{tV} \circ e^{sW} \neq e^{sW} \circ e^{tV}$, moreover, $e^{tV} \circ e^{tW} \neq e^{tW} \circ e^{tV}$. So the curve

$$\varphi(t) = e^{-tW} \circ e^{-tV} \circ e^{tW} \circ e^{tV}(q_0)$$

satisfies the inequality $\varphi(t) \neq q_0, t \in \mathbb{R}$. The leading nontrivial term of the Taylor expansion of $\varphi(t), t \to 0$, is taken as the measure of noncommutativity of vector fields V and W. Namely, we have $\varphi(0) = 0, \dot{\varphi}(0) = 0, \ddot{\varphi}(0) \neq 0$ generically. Thus the *commutator* (*Lie bracket*) of the vector fields V, W at the point q_0 is defined as

$$[V, W](q_0) := \frac{1}{2}\ddot{\varphi}(0),$$

so that

$$\varphi(t) = q_0 + t^2[V, W](q_0) + o(t^2), \qquad t \to 0.$$

In local coordinates

$$[V, W] = \frac{\partial W}{\partial x}V - \frac{\partial V}{\partial x}W. \qquad (1.20)$$

1.2.2.1 Example: Car in the Plane

Consider the vector fields in the right-hand side of the control system of Sect. 1.1.1.4

$$\begin{pmatrix} \dot{x} \\ \dot{y} \\ \dot{\theta} \end{pmatrix} = u \begin{pmatrix} \cos\theta \\ \sin\theta \\ 0 \end{pmatrix} + v \begin{pmatrix} 0 \\ 0 \\ 1 \end{pmatrix},$$

$$V = \cos\theta \frac{\partial}{\partial x} + \sin\theta \frac{\partial}{\partial y}, \qquad W = \frac{\partial}{\partial \theta}.$$

Compute their Lie bracket:

$$[V, W] = \frac{\partial W}{\partial q}V - \frac{\partial V}{\partial q}W = 0 \cdot V - \begin{pmatrix} 0 & 0 & -\sin\theta \\ 0 & 0 & \cos\theta \\ 0 & 0 & 0 \end{pmatrix}\begin{pmatrix} 0 \\ 0 \\ 1 \end{pmatrix} = \begin{pmatrix} \sin\theta \\ -\cos\theta \\ 0 \end{pmatrix}.$$

There is another way of computing Lie brackets, via commutator of differential operators corresponding to vector fields:

$$[V, W] = V \circ W - W \circ V = \left(\cos\theta \frac{\partial}{\partial x} + \sin\theta \frac{\partial}{\partial y}\right)\frac{\partial}{\partial \theta} - \frac{\partial}{\partial \theta}\left(\cos\theta \frac{\partial}{\partial x} + \sin\theta \frac{\partial}{\partial y}\right)$$

$$= \sin\theta \frac{\partial}{\partial x} - \cos\theta \frac{\partial}{\partial y}.$$

Notice the visual meaning of the vector fields $V, W, [V, W]$ for the car in the plane:

- V generates the motion forward
- W generates rotations of the car
- $[V, W]$ generates motion of the car in the direction perpendicular to its orientation, thus physically forbidden.

Choosing alternating motions of the car:

forward $\rightarrow$ rotation counter-clockwise $\rightarrow$ backward $\rightarrow$ rotation clockwise,

we can move the car infinitesimally in the forbidden direction. So the Lie bracket $[V, W]$ is generated by a car during parking manoeuvres in a limited space.

1.2.3 Lie Groups and Lie Algebras

A set G is called a *Lie group* if it is a smooth manifold endowed with a group structure such that the following mappings are smooth:

$$(g, h) \mapsto gh, \qquad G \times G \to G,$$

$$g \mapsto g^{-1}, \qquad G \to G.$$

Let $\mathrm{Id} \in G$ denote the identity element of the group G.

Denote by $\mathbb{R}^{n \times n}$ the set of al real $n \times n$ matrices. The set

$$\mathrm{GL}(n, \mathbb{R}) = \{g \in \mathbb{R}^{n \times n} \mid \det g \neq 0\}$$

is obviously a Lie group w.r.t. the matrix product, it is called the *general linear group*. The main examples of Lie groups are *linear Lie groups*, i.e., closed subgroups of $\mathrm{GL}(n, \mathbb{R})$; see Exercise 6 in Sect. 1.3.

A set $\mathfrak{g}$ is called a *Lie algebra* if it is a vector space endowed with a binary operation $[\cdot, \cdot]$ called *Lie bracket* that satisfies the following properties:

(1) bilinearity: $\qquad [ax+by, z] = a[x, z] + b[y, z], \qquad x, y, z \in \mathfrak{g}, \quad a, b \in \mathbb{R},$
(2) skew symmetry: $\qquad [x, y] = -[y, x], \qquad x, y \in \mathfrak{g},$
(3) Jacobi identity: $\qquad [x, [y, z]] + [y, [z, x]] + [z, [x, y]] = 0, \qquad x, y, z \in \mathfrak{g}.$

For any element g of a Lie group G, the mapping

$$L_g : h \mapsto gh, \qquad G \to G,$$

is called the *left translation* by g. A vector field $X \in \mathrm{Vec}(G)$ is called *left-invariant* if it is preserved by left translations:

$$(L_g)_*(X(h)) = X(gh), \qquad g,\, h \in G.$$

Lie bracket of left-invariant vector fields is left-invariant. Thus left-invariant vector fields on a Lie group G form a Lie algebra $\mathfrak{g}$ called the *Lie algebra of the Lie group G*. Any left-invariant vector field $X \in \mathfrak{g}$ is uniquely determined by its value $X(\mathrm{Id}) \in T_{\mathrm{Id}}G$, and vice versa. So there is a linear isomorphism $\mathfrak{g} \cong T_{\mathrm{Id}}G$, which defines the structure of a Lie algebra on $T_{\mathrm{Id}}G$. Thus the tangent space $T_{\mathrm{Id}}G$ is also called the Lie algebra of the Lie group G.

For a Lie group G, the tangent space at a point $g \in G$ is

$$T_g G = (L_g)_* T_{\mathrm{Id}}G, \qquad g \in G. \tag{1.21}$$

In the case of a linear Lie group $G \subset \mathrm{GL}(n, \mathbb{R})$, the differential of the left translation can be represented as a matrix product:

$$(L_g)_* A = gA, \qquad g \in G, \quad A \in T_{\mathrm{Id}}G.$$

Thus left-invariant vector fields on a linear Lie group G have the form

$$V(g) = gA, \qquad g \in G, \quad A \in T_{\mathrm{Id}}G. \tag{1.22}$$

A control system on a Lie group G

$$\dot{g} = f(g, u), \qquad g \in G, \quad u \in U,$$

is called *left-invariant* if its dynamics is preserved by left translations:

$$(L_h)_* f(g, u) = f(hg, u), \qquad g,\, h \in G, \quad u \in U.$$

Similarly, an optimal control problem on G is called *left-invariant* if both its dynamics and the cost functional are preserved by left translations.

If an optimal control problem is left-invariant on a Lie group, then the initial point of trajectories can be set to the identity element of the group: $g(0) = \mathrm{Id}$, since any trajectory can be mapped by left translations to satisfy such initial condition.

1.3   Exercises

1. Describe the attainable sets $\mathcal{A}_{q_0}$ for the examples of Sects. 1.1.1.1–1.1.1.8. Which of these systems are controllable?

2. Describe in the example of Sect. 1.1.1.6:

$$\text{Lie}_q(X_1, X_2)$$

$$= \text{span}(X_1(q), X_2(q), [X_1, X_2](q), [X_1, [X_1, X_2]](q), [X_2, [X_1, X_2]](q), \dots),$$

where X_1 and X_2 are the vector fields in the right-hand side of system (1.7), (1.8):

$$\dot{q} = u_1 X_1 + u_2 X_2, \qquad q \in \mathbb{R}^2 \times SO(3).$$

3. Show that the two-dimensional sphere S^2 and the group $SO(3)$ of rotations of the 3-space are smooth submanifolds. Compute their tangent spaces. Prove that S^2 is not a Lie group.

4. Prove in example of Sect. 1.1.1.7:

$$l = \int_0^{t_1} \sqrt{u_1^2 + u_2^2} \, dt \to \min \quad \Leftrightarrow \quad J = \int_0^{t_1} (u_1^2 + u_2^2) \, dt \to \min$$

for a fixed terminal time t_1.

5. Prove formula (1.20).

6. Show that the following sets are linear Lie groups:

- the *special linear group*

$$SL(n, \mathbb{R}) = \{g \in GL(n, \mathbb{R}) \mid \det g = 1\},$$

- the *special orthogonal group*

$$SO(n) = \left\{g \in GL(n, \mathbb{R}) \mid \det g = 1, \ g^{-1} = g^T\right\},$$

- the *special Euclidean group*

$$SE(n) = \left\{\begin{pmatrix} A & v \\ 0 & 1 \end{pmatrix} \in GL(n+1, \mathbb{R}) \mid A \in SO(n), \ v \in \mathbb{R}^n\right\},$$

- the *special unitary group*

$$SU(n) = \left\{\begin{pmatrix} A & B \\ -B & A \end{pmatrix} \in \mathbb{R}^{2n \times 2n} \mid A, B \in \mathbb{R}^{n \times n}, \ AA^T + BB^T = \text{Id}, \right.$$

$$\left. BA^T - AB^T = 0, \ \det(A + iB) = 1\right\},$$

compute their dimensions.

7. Compute the Lie algebras of the Lie groups of the previous item as their tangent spaces at the identity elements.
8. Prove formula (1.21).
9. Prove that left-invariant vector fields on a linear Lie group G have the form (1.22).
10. Show that in each example of Sects. 1.1.1.1–1.1.1.8 the state space has a natural structure of a Lie group. Which of these examples give left-invariant optimal control problems?

Chapter 2
Controllability Problem

On a yonder branch perches a nightingale cheerfully singing;
The sun is warm, and a soothing breeze blows, on the bank the willows are green;
The ox is there all by himself, nowhere is he to hide himself;
The splendid head decorated with stately horns what painter can reproduce him?

Pu-ming, "The Ten Oxherding Pictures" (cited by Suzuki [110])

Seeing the Ox

In this chapter we study the controllability problem. First we prove the classic Kalman controllability test for linear autonomous systems and a related sufficient local controllability condition for nonlinear systems via linearisation. Then we prove the fundamental Nagano–Sussmann Orbit theorem and its corollaries, including

Yu. Sachkov, *Introduction to Geometric Control*, Springer Optimization
and Its Applications 192, https://doi.org/10.1007/978-3-031-02070-4_2

the Rashevskii–Chow and the Frobenius theorems. Finally, we prove an important Krener's theorem on attainable sets of full-rank systems. Theoretical development is illustrated by the study of systems given in Sect. 1.1.1.

2.1 Controllability

In this section we prove some basic facts on the controllability problem for linear and nonlinear systems.

2.1.1 Controllability of Linear Systems

We start from the simplest class of control systems, quite popular in applications. *Linear control systems* have the form

$$\dot{x} = Ax + \sum_{i=1}^{k} u_i b_i = Ax + Bu, \tag{2.1}$$

$$x \in \mathbb{R}^n, \quad u = (u_1, \ldots, u_k) \in \mathbb{R}^k, \quad u(\cdot) \in L^1([0, t_1], \mathbb{R}^k).$$

Here A and $B = (b_1, \ldots, b_k)$ are constant $n \times n$ and $n \times k$ matrices respectively, $b_1, \ldots, b_k \in \mathbb{R}^n$.

It is easy to find solutions to such systems by the variation of constants method. The linear ODE $\dot{x} = Ax$ has a solution

$$x(t) = e^{At} C, \qquad C \equiv \text{const} \in \mathbb{R}^n,$$

where

$$e^{At} = \sum_{k=0}^{\infty} \frac{(At)^k}{k!}$$

is the matrix exponential. Let us look for solutions to system (2.1) in the form $x(t) = e^{At} C(t)$. We substitute this expression into (2.1) and get

$$\dot{x} = Ae^{At} C + e^{At} \dot{C} = Ae^{At} C + Bu,$$

$$\dot{C}(t) = e^{-At} Bu(t),$$

$$C(t) = \int_0^t e^{-As} Bu(s) \, ds + C_0,$$

$$x(t) = e^{At} \left(\int_0^t e^{-As} Bu(s)\, ds + C_0 \right),$$

$$x(0) = C_0 = x_0,$$

$$x(t) = e^{At} \left(x_0 + \int_0^t e^{-As} Bu(s)\, ds \right). \qquad (2.2)$$

Formula (2.2) is called *Cauchy's formula* for linear systems.

We use Cauchy's formula to prove the classic *Kalman controllability test* for linear systems.

Definition 2.1 A linear system (2.1) is called *controllable* from a point $x_0 \in \mathbb{R}^n$ for time $t_1 > 0$ (for time not greater than t_1) if

$$\mathcal{A}_{x_0}(t_1) = \mathbb{R}^n \qquad (\text{resp. } \mathcal{A}_{x_0}(\le t_1) = \mathbb{R}^n).$$

Theorem 2.1 (Kalman Controllability Test) Let $t_1 > 0$ and $x_0 \in \mathbb{R}^n$. A linear system (2.1) is controllable from x_0 for time t_1 iff

$$\operatorname{span}(B, AB, \ldots, A^{n-1}B) = \mathbb{R}^n. \qquad (2.3)$$

Proof The mapping $L^1 \ni u(\cdot) \mapsto x(t_1) \in \mathbb{R}^n$ is affine, thus its image $\mathcal{A}_{x_0}(t_1)$ is an affine subspace of $\mathbb{R}^n$. Further we rewrite the definition of controllability taking into account Cauchy's formula (2.2):

$$\mathcal{A}_{x_0}(t_1) = \mathbb{R}^n \Leftrightarrow \operatorname{Im} e^{At_1} \left(x_0 + \int_0^{t_1} e^{-At} Bu(t)\, dt \right) = \mathbb{R}^n$$

$$\Leftrightarrow \operatorname{Im} \int_0^{t_1} e^{-At} Bu(t)\, dt = \mathbb{R}^n.$$

Now we prove the necessity. Let $\mathcal{A}_{x_0}(t_1) = \mathbb{R}^n$, but $\operatorname{span}(B, AB, \ldots, A^{n-1}B) \ne \mathbb{R}^n$. Then there exists a covector $0 \ne p \in \mathbb{R}^{n*}$ such that

$$pA^i B = 0, \quad i = 0, \ldots, n-1.$$

By the Cayley–Hamilton theorem, $A^n = \sum_{i=0}^{n-1} \alpha_i A^i$ for some $\alpha_i \in \mathbb{R}$. Thus

$$A^m = \sum_{i=0}^{n-1} \beta_i^m A^i, \quad \beta_i^m \in \mathbb{R}, \quad m = 0, 1, 2, \ldots.$$

Consequently,

$$pA^m B = \sum_{i=0}^{n-1} \beta_i^m pA^i B = 0, \qquad m = 0, 1, 2, \ldots,$$

$$pe^{-At} B = p \sum_{m=0}^{\infty} \frac{(-At)^m}{m!} B = 0,$$

and $\text{Im} \int_0^{t_1} e^{-At} Bu(t)\, dt \neq \mathbb{R}^n$, a contradiction.

Then we prove the sufficiency. Let $\text{span}(B, AB, \ldots, A^{n-1}B) = \mathbb{R}^n$, but

$$\text{Im} \int_0^{t_1} e^{-At} Bu(t)\, dt \neq \mathbb{R}^n.$$

Then there exists a covector $0 \neq p \in \mathbb{R}^{n*}$ such that

$$p \int_0^{t_1} e^{-At} Bu(t)\, dt = 0 \qquad \forall u \in L^1([0, t_1], \mathbb{R}^k).$$

Let $e_1, \ldots, e_k$ be the standard frame in $\mathbb{R}^k$. For any $\tau \in [0, t_1]$ and any $i = 1, \ldots, k$, define the following controls:

$$u(t) = \begin{cases} e_i, & t \in [0, \tau], \\ 0, & t \in (\tau, t_1]. \end{cases} \tag{2.4}$$

We have

$$\int_0^{t_1} e^{-At} Bu(t)\, dt = \int_0^{\tau} e^{-At} b_i\, dt = \frac{\text{Id} - e^{-A\tau}}{A} b_i,$$

thus

$$p \frac{\text{Id} - e^{-A\tau}}{A} B = 0, \tag{2.5}$$

where

$$\frac{\text{Id} - e^{-A\tau}}{A} = \sum_{m=1}^{\infty} (-1)^{m-1} \frac{\tau^m}{m!} A^{m-1}.$$

We differentiate successively identity (2.5) at $\tau = 0$ and obtain

$$pB = pAB = \cdots = pA^{n-1}B = 0,$$

thus span$(B, AB, \ldots, A^{n-1}B) \neq \mathbb{R}^n$, a contradiction. □

Condition (2.3) is called the *Kalman controllability condition*.

Remark 2.1 Control (2.4) is piecewise constant. Thus if Kalman's condition (2.3) holds, then linear system (2.1) is controllable for any time $t_1 > 0$ with piecewise-constant controls.

For linear systems, controllability for the class of admissible controls $u(\cdot) \in L^1$ is equivalent to controllability for any class of admissible controls $u(\cdot) \in L$ where L is a linear subspace of L^1 containing piecewise constant functions.

Corollary 2.1 *The following conditions are equivalent:*

- *the Kalman controllability condition (2.3)*
- $\forall t_1 > 0 \, \forall x_0 \in \mathbb{R}^n$ *linear system (2.1) is controllable from x_0 for time t_1*
- $\forall t_1 > 0 \, \forall x_0 \in \mathbb{R}^n$ *linear system (2.1) is controllable from x_0 for time not greater than t_1*
- $\exists t_1 > 0 \, \exists x_0 \in \mathbb{R}^n$ *such that linear system (2.1) is controllable from x_0 for time t_1*
- $\exists t_1 > 0 \, \exists x_0 \in \mathbb{R}^n$ *such that linear system (2.1) is controllable from x_0 for time not greater than t_1.*

In these cases linear system (2.1) is called *controllable*.

2.1.2 Local Controllability of Nonlinear Systems

Consider now a nonlinear system

$$\dot{x} = f(x, u), \qquad x \in \mathbb{R}^n, \quad u \in U \subset \mathbb{R}^m, \tag{2.6}$$

where the right-hand side $f : \mathbb{R}^n \times U \to \mathbb{R}^n$ is smooth. Admissible controls are Lebesgue measurable bounded mappings $u(\cdot) \in L^\infty([0, t_1], U)$.

A point $(x_0, u_0) \in \mathbb{R}^n \times U$ is called an *equilibrium point* of system (2.6) if $f(x_0, u_0) = 0$. We will suppose that

$$u_0 \in \text{int } U \tag{2.7}$$

and consider the *linearisation* of system (2.6) at the equilibrium point (x_0, u_0):

$$\dot{y} = Ay + Bv, \qquad y \in \mathbb{R}^n, \quad v \in \mathbb{R}^m, \tag{2.8}$$

$$A = \frac{\partial f}{\partial x}\bigg|_{(x_0, u_0)}, \quad B = \frac{\partial f}{\partial u}\bigg|_{(x_0, u_0)}.$$

It is natural to expect that global properties of linearisation (2.8) imply the corresponding local properties of nonlinear system (2.6). Indeed, there holds the following statement.

Theorem 2.2 (Linearisation Principle for Controllability) *If linearisation* (2.8) *is controllable at an equilibrium point* (x_0, u_0) *with* (2.7), *then for any* $t_1 > 0$ *nonlinear system* (2.6) *is locally controllable at the point* x_0 *for time* t_1:

$$\forall t_1 > 0 \quad x_0 \in \operatorname{int} \mathcal{A}_{x_0}(t_1).$$

Thus nonlinear system (2.6) *is STLC at* x_0.

Proof Fix any $t_1 > 0$. Let $e_1, \ldots, e_n$ be the standard frame in $\mathbb{R}^n$. Since linear system (2.8) is controllable, then

$$\forall i = 1, \ldots, n \quad \exists v_i \in L^\infty([0, t_1], \mathbb{R}^m) : \quad y_{v_i}(0) = 0, \quad y_{v_i}(t_1) = e_i. \tag{2.9}$$

Construct the following family of controls:

$$u(z, t) = u_0 + z_1 v_1(t) + \cdots + z_n v_n(t), \quad z = (z_1, \ldots, z_n) \in \mathbb{R}^n.$$

By condition (2.7), for sufficiently small $|z|$ and any $t \in [0, t_1]$, the control $u(z, t) \in U$, thus it is admissible for nonlinear system (2.6). Consider the corresponding family of trajectories of (2.6):

$$x(z, t) = x_{u(z, \cdot)}(t), \quad x(z, 0) = x_0, \quad z \in B,$$

where B is a small open ball in $\mathbb{R}^n$ centred at the origin. Since

$$x(z, t_1) \in \mathcal{A}_{x_0}(t_1), \quad z \in B,$$

then the mapping

$$F : z \mapsto x(z, t_1), \quad B \to \mathbb{R}^n$$

satisfies the inclusion

$$F(B) \subset \mathcal{A}_{x_0}(t_1).$$

It remains to show that $x_0 \in \operatorname{int} F(B)$. To this end define the matrix function

$$W(t) = \left. \frac{\partial x(z, t)}{\partial z} \right|_{z=0}.$$

We show that $\det W(t_1) = \left. \frac{\partial F}{\partial z} \right|_{z=0} \neq 0$. This would imply that

$$x_0 = F(0) \in \operatorname{int} F(B) \subset \mathcal{A}_{x_0}(t_1).$$

Differentiating the identity $\frac{\partial x}{\partial t} = f(x, u(z, t))$ w.r.t. z, we get

$$\frac{\partial}{\partial t} \frac{\partial x}{\partial z}\bigg|_{z=0} = \frac{\partial f}{\partial x}\bigg|_{(x_0, u_0)} \frac{\partial x}{\partial z}\bigg|_{z=0} + \frac{\partial f}{\partial u}\bigg|_{(x_0, u_0)} \frac{\partial u}{\partial z}\bigg|_{z=0}$$

since $u(0, t) \equiv u_0$ and $x(0, t) \equiv x_0$. Thus we get a matrix ODE

$$\dot{W}(t) = AW(t) + B(v_1(t), \ldots, v_n(t)) \tag{2.10}$$

with the initial condition

$$W(0) = \frac{\partial x(z, 0)}{\partial z}\bigg|_{z=0} = \frac{\partial x_0}{\partial z}\bigg|_{z=0} = 0.$$

ODE (2.10) means that columns of the matrix $W(t)$ are solutions to linear system (2.8) with the control $v_i(t)$. By condition (2.9) we have $W(t_1) = (e_1, \ldots, e_n)$, so $\det W(t_1) = 1 \neq 0$.

By the implicit function theorem, we have $x_0 \in \text{int } F(B)$, thus $x_0 \in \text{int } \mathcal{A}_{x_0}(t_1)$.

$\square$

2.1.2.1 Example

Consider a control system

$$\dot{x} = uf_1(x) + (1 - u)f_2(x), \qquad x = (x_1, x_2) \in \mathbb{R}^2, \quad u \in [0, 1], \tag{2.11}$$

$$f_1(x) = \frac{\partial}{\partial x_1}, \qquad f_2(x) = -\frac{\partial}{\partial x_1} + x_1\frac{\partial}{\partial x_2}. \tag{2.12}$$

The point $(x^0, u^0) = (0, \frac{1}{2})$ is an equilibrium point of the system, moreover, $u^0 \in \text{int}([0, 1])$. The linearisation of system (2.11) at the equilibrium point (x^0, u^0) has the form

$$\dot{y} = Ay + Bv, \qquad y \in \mathbb{R}^2, \quad v \in \mathbb{R}, \tag{2.13}$$

where

$$A = \begin{pmatrix} 0 & 0 \\ \frac{1}{2} & 0 \end{pmatrix}, \qquad B = \begin{pmatrix} 2 \\ 0 \end{pmatrix}.$$

Let us check Kalman's condition for the linearisation:

$$\text{rank}(B, AB) = \text{rank}\begin{pmatrix} 2 & 0 \\ 0 & 1 \end{pmatrix} = 2,$$

thus linear system (2.13) is controllable. So nonlinear system (2.11) is locally controllable at the point x^0 for any time $t_1 > 0$.

2.1.3   Exercises

1. For the sub-Riemannian problem on the group of motions of the plane, find equilibrium points and study controllability of linearisation at these points.
2. For Euler's elastic problem, find equilibrium points and study controllability of linearisation at these points.
3. Prove local and global controllability of system (2.11), (2.12) geometrically, with the help of the phase portraits of the vector fields f_1, f_2.

2.2   The Orbit Theorem

This section is devoted to a fundamental geometric control theorem on an orbit of a control system.

2.2.1   Orbit of a Control System

In this section a *control system* on a smooth manifold M is an arbitrary set of vector fields $\mathcal{F} \subset \mathrm{Vec}(M)$. We assume for simplicity that all vector fields in $\mathcal{F}$ are *complete*, i.e., have trajectories defined for any real time. The *attainable set* of the system $\mathcal{F}$ from a point $q_0 \in M$ is defined as

$$\mathcal{A}_{q_0} = \{ e^{t_N f_N} \circ \cdots \circ e^{t_1 f_1}(q_0) \mid t_i \geq 0, \quad f_i \in \mathcal{F}, \quad N \in \mathbb{N} \}.$$

If we parametrize $\mathcal{F}$ by a control parameter u, such attainable set corresponds to piecewise constant controls and arbitrary nonnegative times.

Before studying the attainable set, in which only the forward motion is possible, we consider a bigger set, in which one can move both forward and backward—the *orbit* of the system $\mathcal{F}$ through the point q_0:

$$O_{q_0} = \{ e^{t_N f_N} \circ \cdots \circ e^{t_1 f_1}(q_0) \mid t_i \in \mathbb{R}, \quad f_i \in \mathcal{F}, \quad N \in \mathbb{N} \}.$$

There hold the following relations between attainable sets and orbits:

1. $\mathcal{A}_{q_0} \subset O_{q_0}$, this inclusion is obvious
2. O_{q_0} has a "simpler" structure than $\mathcal{A}_{q_0}$
3. $\mathcal{A}_{q_0}$ has a "reasonable" structure inside O_{q_0}.

We clarify relations 2, 3 in the Orbit theorem (Theorem 2.3) and in Krener's theorem (Theorem 2.6) respectively.

A system $\mathcal{F}$ is called *symmetric* if $\mathcal{F} = -\mathcal{F}$. The following property is obvious:

4. $\mathcal{F} = -\mathcal{F} \quad \Rightarrow \quad \mathcal{A}_{q_0} = O_{q_0}.$

2.2.2 Preliminaries

Before stating the Orbit theorem we recall some necessary facts about action of diffeomorphisms on vector fields and about immersed submanifolds.

2.2.2.1 Action of Diffeomorphisms on Tangent Vectors and Vector Fields

Let M, N be smooth manifolds, $q \in M$, and let $v \in T_q M$ be a tangent vector. Let $\Phi \colon M \to N$ be a smooth mapping. Then the action (*push-forward*) of the mapping Φ on the vector v is defined as follows. Let $\varphi \colon (-\varepsilon, \varepsilon) \to M$ be a smooth curve such that $\varphi(0) = q$, $\dot{\varphi}(0) = v$. Then the tangent vector $\Phi_{*q} v \in T_{\Phi(q)} N$ is defined as $\Phi_{*q} v = \frac{d}{dt}\big|_{t=0} \Phi \circ \varphi(t)$. Thus we have a mapping $D_q \Phi = \Phi_{*q} \colon T_q M \to T_{\Phi(q)} N$, the *differential* of the mapping Φ at the point q.

Now let $V \in \operatorname{Vec}(M)$ be a smooth vector field, and let $\Phi \colon M \to N$ be a *diffeomorphism*, i.e., a smooth bijective mapping with a smooth inverse. Then the vector field $\Phi_* V \in \operatorname{Vec}(N)$ is defined by the equality

$$\Phi_* V|_{\Phi(q)} = \frac{d}{dt}\bigg|_{t=0} \Phi \circ e^{tV}(q) = \Phi_{*q}(V(q)).$$

Thus we have a mapping

$$\Phi_* \colon \operatorname{Vec}(M) \to \operatorname{Vec}(N),$$

push-forward of vector fields from the manifold M to the manifold N under the action of the diffeomorphism Φ.

In the proof of the Orbit theorem we will need the following obvious property related to action of diffeomorphisms on vector fields and tangent vectors.

Remark 2.2 Let $\Phi \colon M \to N$ be a diffeomorphism between manifolds, and let $V \in \operatorname{Vec}(M)$, $q \in M$. Then

$$\Phi_{*q}(V(q)) = (\Phi_* V)(\Phi(q)).$$

This equality follows immediately from the definitions of the action of diffeomorphisms on vector fields and tangent vectors.

2.2.2.2   Immersed Submanifolds

Definition 2.2 A subset W of a smooth manifold M is called a k-dimensional *immersed submanifold* of M if there exists a k-dimensional manifold N and a smooth mapping $F \colon N \to M$ such that:

- F is injective
- $\text{Ker } F_{*q} = 0$ for any $q \in N$
- $W = F(N)$.

Example 2.1 Figure of eight: prove that the curve

$$\left\{ x = \sin 2\varphi \cos^2 \varphi, \quad y = \sin 2\varphi \sin^2 \varphi \mid \varphi \in (0, \pi) \right\}$$

is a 1-dimensional immersed submanifold of the 2-dimensional plane; see Fig. 2.1.

Example 2.2 Consider the two-dimensional torus

$$\mathbb{T}^2 = \mathbb{R}^2 / (2\pi \, \mathbb{Z}^2) = \{(x, y) \in S^1 \times S^1\},$$

and consider a vector field on it with constant coefficients: $V = p\frac{\partial}{\partial x} + q\frac{\partial}{\partial y} \in \text{Vec}(\mathbb{T}^2)$, $p^2 + q^2 \neq 0$. The orbit O_0 of the vector field V through the origin $0 \in \mathbb{T}^2$ may have two different qualitative types:

(1) $p/q \in \mathbb{Q} \cup \{\infty\}$. Then the orbit of V is closed: $\text{cl } O_0 = O_0$.
(2) $p/q \in \mathbb{R} \backslash \mathbb{Q}$. Then the orbit is dense in the torus: $\text{cl } O_0 = \mathbb{T}^2$. In this case the orbit O_0 is called the *irrational winding of the torus*.

(Here and below cl denotes closure of a set.) In the both cases the orbit O_0 is an immersed submanifold of the torus, but in the second case it is not embedded.

So even for one vector field the orbit may be an immersed submanifold, but not an embedded one (an immersed submanifold $N = F(W) \subset M$ is called *embedded* if $F : W \to N$ is a homeomorphism in the topology induced by the inclusion

Fig. 2.1 Immersed submanifold

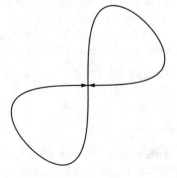

$N \subset M$). In case (2) the topology of the orbit induced by the inclusion $O_0 \subset \mathbb{R}^2$ is weaker than the topology of the orbit induced by the immersion

$$t \mapsto e^{tV}(0), \quad \mathbb{R} \to O_0.$$

2.2.3 The Orbit Theorem

Now we can state and prove the following fundamental *Orbit theorem*.

Theorem 2.3 (Orbit Theorem, Nagano–Sussmann) *Let $\mathcal{F} \subset \mathrm{Vec}(M)$, and let $q_0 \in M$.*

(1) *The orbit O_{q_0} is a connected immersed submanifold of M.*
(2) *For any $q \in O_{q_0}$*

$$T_q O_{q_0} = \mathrm{span}(\mathcal{P}_*\mathcal{F})(q) = \mathrm{span}\{(P_*V)(q) \mid P \in \mathcal{P}, \quad V \in \mathcal{F}\}, \qquad (2.14)$$

$$\mathcal{P} = \{e^{t_N f_N} \circ \cdots \circ e^{t_1 f_1} \mid t_i \in \mathbb{R}, \quad f_i \in \mathcal{F}, \quad N \in \mathbb{N}\}.$$

Proof Introduce a vector space important in the sequel

$$\Pi_q = \mathrm{span}(\mathcal{P}_*\mathcal{F})(q) \subset T_q M, \qquad q \in M,$$

this is a candidate tangent space to the orbit O_{q_0}; see (2.14).

(1) We prove that for all $q \in O_{q_0}$

$$\dim \Pi_q = \dim \Pi_{q_0}.$$

Choose any point $q \in O_{q_0}$, then $q = Q(q_0)$, $Q \in \mathcal{P}$. Let us show that $Q_*^{-1}(\Pi_q) \subset \Pi_{q_0}$.

Choose any element $(P_*f)(q) \in \Pi_q$, $P \in \mathcal{P}$, $f \in \mathcal{F}$. Then

$$Q_*^{-1}[(P_*f)(q)] = (Q_*^{-1} \circ P_*f)(Q^{-1}(q))$$

$$= [(Q^{-1} \circ P)_* f](q_0) \in (\mathcal{P}_*\mathcal{F})(q_0) \subset \Pi_{q_0}.$$

Thus $Q_*^{-1}(\Pi_q) \subset \Pi_{q_0}$, whence

$$\dim \Pi_q \leq \dim \Pi_{q_0}.$$

Interchanging in this arguments q and q_0, we get $\dim \Pi_{q_0} \leq \dim \Pi_q$.

Finally we have $\dim \Pi_q = \dim \Pi_{q_0}$, $q \in O_{q_0}$.

(2) For any point $q \in M$ denote $m = \dim \Pi_q$, and choose such vector fields $V_1, \ldots, V_m \in \mathcal{P}_* \mathcal{F}$ that

$$\Pi_q = \mathrm{span}(V_1(q), \ldots, V_m(q)).$$

Further, define a mapping

$$G_q : (t_1, \ldots, t_m) \mapsto e^{t_m V_m} \circ \cdots \circ e^{t_1 V_1}(q), \qquad \mathbb{R}^m \to M.$$

We have $\frac{\partial G_q}{\partial t_i}(0) = V_i(q)$, thus the vectors $\frac{\partial G_q}{\partial t_1}(0), \ldots, \frac{\partial G_q}{\partial t_m}(0)$ are linearly independent.

Consequently, the restriction of G_q to a sufficiently small neighbourhood W_0 of the origin in $\mathbb{R}^m$ is a submersion.

(3) The image $G_q(W_0)$ is an (embedded) submanifold of M, may be, for a smaller neighbourhood W_0.

(4) We show that $G_q(W_0) \subset O_q$.
    We have $G_q(W_0) = \{e^{t_m V_m} \circ \cdots \circ e^{t_1 V_1}(q) \mid t \in W_0\}$.
    Since $V_1 = P_* f$, $P \in \mathcal{P}$, $f \in \mathcal{F}$, we get

$$e^{t_1 V_1}(q) = e^{t_1 P_* f}(q) = P \circ e^{t_1 f} \circ P^{-1}(q) \in O_q.$$

We conclude similarly that $e^{t_2 V_2} \circ e^{t_1 V_1}(q) \in O_q$ etc. Finally we have $G_q(t) \in O_q, t \in W_0$, q.e.d.

(5) We show that $G_{q_*}(T_t \mathbb{R}^m) = \Pi_{G_q(t)}, t \in W_0$. We have $\dim G_{q_*}(T_t \mathbb{R}^m) = m = \dim \Pi_{G_q(t)}$, thus it suffices to prove the inclusion

$$\frac{\partial G_q}{\partial t_i}(t) \in \Pi_{G_q(t)}, \qquad t \in W_0.$$

Let us compute this partial derivative:

$$\frac{\partial G_q}{\partial t_i} = \frac{\partial}{\partial t_i} e^{t_m V_m} \circ \cdots \circ e^{t_i V_i} \circ \cdots \circ e^{t_1 V_1}(q)$$

denote $R = e^{t_m V_m} \circ \cdots \circ e^{t_{i+1} V_{i+1}}$, $q' = e^{t_{i-1} V_{i-1}} \circ \cdots \circ e^{t_1 V_1}(q)$,

$$= \frac{\partial}{\partial t_i} R \circ e^{t_i V_i}(q') = R_* V_i(e^{t_i V_i}(q'))$$

$$= (R_* V_i)[R \circ e^{t_i V_i} \circ \cdots \circ e^{t_1 V_1}(q)]$$

$$= (R_* V_i)(G_q(t)) \in (\mathcal{P}_* \mathcal{F})(G_q(t))$$

$$\subset \Pi_{G_q(t)}.$$

Thus $G_{q_*}(T_t\mathbb{R}^m) = \Pi_{G_q(t)}$, i.e., the space $\Pi_{G_q(t)}$ is a tangent space to the smooth manifold $G_q(W_0)$ at the point $G_q(t)$.

(6) We prove that the sets $G_q(W_0)$ form a base of a ("strong") topology on M.

(6a) It is obvious that any point $q \in M$ is contained in the set $G_q(W_0)$.

(6b) Let us show that for any point $\widehat{q} \in G_q(W_0) \cap G_{\widetilde{q}}(\widetilde{W}_0)$ there exists a set $G_{\widehat{q}}(\widehat{W}_0) \subset G_q(W_0) \cap G_{\widetilde{q}}(\widetilde{W}_0)$.

Take any point $\widehat{q} \in G_q(W_0) \cap G_{\widetilde{q}}(\widetilde{W}_0)$ and consider $G_{\widehat{q}}(t) = e^{t_m \widehat{V}_m} \circ \cdots \circ e^{t_1 \widehat{V}_1}(\widehat{q})$. For any point $q' \in G_q(W_0)$ we have $\widehat{V}_1(q') \in (\mathcal{P}_*\mathcal{F})(q') \subset \Pi_{q'}$. But $G_q(W_0)$ is a submanifold with the tangent space $T_{q'}G_q(W_0) = \Pi_{q'}$. The vector field $\widehat{V}_1$ is tangent to this submanifold, thus $e^{t_1 \widehat{V}_1}(\widehat{q}) \in G_q(W_0)$ for small $|t_1|$. We conclude similarly that $e^{t_2 \widehat{V}_2} \circ e^{t_1 \widehat{V}_1}(\widehat{q}) \in G_q(W_0)$ for small $|t_1|$, $|t_2|$ etc. Finally we get

$$G_{\widehat{q}}(t) \in G_q(W_0) \text{ for small } |t|.$$

Similarly

$$G_{\widehat{q}}(t) \in G_{\widetilde{q}}(\widetilde{W}_0) \text{ for small } |t|.$$

Thus $G_{\widehat{q}}(\widehat{W}_0) \subset G_q(W_0) \cap G_{\widetilde{q}}(\widetilde{W}_0)$ for some neighbourhood $\widehat{W}_0$, and property (6b) is proved.

It follows from properties (6a) and (6b) that the sets $G_q(W_0)$ form a base of topology on the set M. Denote the corresponding topological space as $M^{\mathcal{F}}$.

(7) We show that for any $q_0 \in M$ the orbit O_{q_0} is connected, open and closed in the space $M^{\mathcal{F}}$.

The mappings $t_i \mapsto e^{t_i f_i}(q)$ are continuous in $M^{\mathcal{F}}$ (prove!), thus O_{q_0} is connected.

Any point $q \in O_{q_0}$ is contained in the neighbourhood $G_q(W_0) \subset O_q = O_{q_0}$, thus the orbit is open in $M^{\mathcal{F}}$.

Finally, any orbit is a complement in M to orbits with which it does not intersect. Thus any orbit is closed in $M^{\mathcal{F}}$.

So any orbit O_{q_0} is a connected component of the topological space $M^{\mathcal{F}}$.

(8) Introduce a smooth structure on O_{q_0} as follows:

- the sets $G_q(W_0)$ are called coordinate neighbourhoods
- the mappings $G_q^{-1} : G_q(W_0) \to W_0$ are called coordinate mappings.

It is easy to see that these coordinate neighbourhoods and mappings agree: for any intersecting neighbourhoods $G_q(W_0)$ and $G_{\widetilde{q}}(\widetilde{W}_0)$ the composition

$$G_{\widetilde{q}} \circ G_q : G_q^{-1}(G_q(W_0) \cap G_{\widetilde{q}}(\widetilde{W}_0)) \to G_{\widetilde{q}}^{-1}(G_q(W_0) \cap G_{\widetilde{q}}(\widetilde{W}_0))$$

is a diffeomorphism (prove!). Thus the orbit O_{q_0} is a smooth manifold.

Moreover, $O_{q_0} \subset M$ is an immersed submanifold of dimension $m = \dim \Pi_{q_0}$.

(9) It follows from item (5) above that the smooth manifold O_{q_0} has a tangent space

$$T_q O_{q_0} = \Pi_q = \text{span}(\mathcal{P}_* \mathcal{F})(q), \qquad q \in O_{q_0}.$$

The Orbit theorem is proved.   □

2.2.4 Corollaries of the Orbit Theorem

Now we prove several important corollaries of the Orbit theorem.

Corollary 2.2 *For any $q_0 \in M$ and any $q \in O_{q_0}$*

$$\text{Lie}_q(\mathcal{F}) \subset T_q O_{q_0}, \tag{2.15}$$

where

$$\text{Lie}_q(\mathcal{F}) = \text{span}\{[f_N, [\ldots, [f_2, f_1] \ldots]](q) \mid f_i \in \mathcal{F}, \ N \in \mathbb{N}\} \subset T_q M.$$

Proof Let $q_0 \in M, q \in O_{q_0}$. Take any $f \in \mathcal{F}$. Then $\varphi(t) = e^{tf}(q) \in O_{q_0}$, thus

$$\dot{\varphi}(0) = f(q) \in T_q O_{q_0}.$$

It follows that $\mathcal{F}(q) \subset T_q O_{q_0}$.

Further, take any $f_1, f_2 \in \mathcal{F}$, then $\varphi(t) = e^{-tf_2} \circ e^{-tf_1} \circ e^{tf_2} \circ e^{tf_1}(q) \in O_{q_0}$. Thus

$$\frac{d}{dt}\Big|_{t=0} \varphi(\sqrt{t}) = [f_1, f_2](q) \in T_q O_{q_0}.$$

It follows that $[\mathcal{F}, \mathcal{F}](q) \subset T_q O_{q_0}$.

We prove similarly that $[[\mathcal{F}, \mathcal{F}], \mathcal{F}](q) \subset T_q O_{q_0}$, and by induction that $\text{Lie}_q(\mathcal{F}) \subset T_q O_{q_0}$.   □

In the analytic case inclusion (2.15) turns into an equality.

Proposition 2.1 *Let M and $\mathcal{F}$ be real-analytic. Then for any $q_0 \in M$ and any $q \in O_{q_0}$*

$$\text{Lie}_q(\mathcal{F}) = T_q O_{q_0}.$$

This proposition is proved in [3]. But in a smooth non-analytic case inclusion (2.15) may become strict.

Example 2.3 Orbit of non-analytic system: let $M = \mathbb{R}^2_{x,y}$, $\mathcal{F} = \{f_1, f_2\}$, $f_1 = \frac{\partial}{\partial x}$, $f_2 = a(x)\frac{\partial}{\partial y}$, where $a \in C^\infty(\mathbb{R})$, $a(x) = 0$ for $x \le 0$, $a(x) > 0$ for $x > 0$.

It is easy to see that $O_q = \mathbb{R}^2$ for any $q = (x, y) \in \mathbb{R}^2$. Although, for $x \le 0$ we have

$$\mathrm{Lie}_q(\mathcal{F}) = \mathrm{span}(f_1(q)) \ne T_q O_q.$$

A system $\mathcal{F} \subset \mathrm{Vec}(M)$ is called *completely nonholonomic* (*full-rank, bracket-generating*) if

$$\mathrm{Lie}_q(\mathcal{F}) = T_q M \quad \forall q \in M.$$

Theorem 2.4 (Rashevskii–Chow) *If* $\mathcal{F} \subset \mathrm{Vec}(M)$ *is completely nonholonomic and M is connected, then $O_q = M$ for any $q \in M$.*

Proof Take any $q \in M$ and any $q_1 \in O_q$. We have $T_{q_1} O_q \supset \mathrm{Lie}_{q_1}(\mathcal{F}) = T_{q_1} M$, thus $\dim O_q = \dim M$, i.e., O_q is open in M.

On the other hand, any orbit is closed as a complement to the union of all other orbits.

Thus any orbit is a connected component of M. Since M is connected, each orbit coincides with M. $\qquad\square$

Corollary 2.3 (Lie Algebra Rank Condition, LARC) *If a manifold M is connected, and a system $\mathcal{F} \subset \mathrm{Vec}(M)$ is symmetric and completely nonholonomic, then it is controllable on M.*

2.2.5 The Frobenius Theorem

A *distribution* on a smooth manifold M is a smooth mapping

$$\Delta: q \mapsto \Delta_q \subset T_q M, \quad q \in M,$$

where the vector subspaces Δ_q have the same dimension called the *rank* of Δ.

An immersed submanifold $N \subset M$ is called an *integral manifold* of a distribution Δ if

$$\forall q \in N \quad T_q N = \Delta_q.$$

A distribution Δ on M is called *integrable* if for any point $q \in M$ there exists an integral manifold $N_q \ni q$.

Denote by

$$\bar{\Delta} = \{f \in \mathrm{Vec}(M) \mid f(q) \in \Delta_q \quad \forall q \in M\}$$

the set of vector fields tangent to Δ.

A distribution Δ is called *holonomic* if $[\bar{\Delta}, \bar{\Delta}] \subset \bar{\Delta}$.

Theorem 2.5 (Frobenius) *A distribution is integrable iff it is holonomic.*

Proof Necessity. Take any $f, g \in \bar{\Delta}$. Let $q \in M$, and let $N_q \ni q$ be the integral manifold of Δ through q. Then

$$\varphi(t) = e^{-tg} \circ e^{-tf} \circ e^{tg} \circ e^{tf}(q) \in N_q,$$

thus

$$\left.\frac{d}{dt}\right|_{t=0} \varphi(\sqrt{t}) = [f, g](q) \in T_q N_q = \Delta_q.$$

So $[f, g] \in \bar{\Delta}$, and the inclusion $[\bar{\Delta}, \bar{\Delta}] \subset \bar{\Delta}$ follows.

Sufficiency. We consider only the analytic case. We have

$$[\bar{\Delta}, \bar{\Delta}] \subset \bar{\Delta}, \qquad [[\bar{\Delta}, \bar{\Delta}], \bar{\Delta}] \subset [\bar{\Delta}, \bar{\Delta}] \subset \bar{\Delta},$$

and inductively $\mathrm{Lie}_q(\bar{\Delta}) \subset \bar{\Delta}_q = \Delta_q$. The reverse inclusion is obvious, thus $\mathrm{Lie}_q(\bar{\Delta}) = \Delta_q$, $q \in M$. Denote $N_q = O_q(\bar{\Delta})$ and prove that N_q is an integral manifold of Δ:

$$T_{q'} N_q = T_{q'}(O_q(\bar{\Delta})) = \mathrm{Lie}_{q'}(\bar{\Delta}) = \Delta_{q'}, \quad q' \in N_q.$$

So $N_q \ni q$ is the integral manifold of Δ, and Δ is integrable. $\qquad\qquad\square$

Consider a *local frame* of Δ:

$$\Delta_q = \mathrm{span}(f_1(q), \ldots, f_k(q)), \quad q \in S \subset M, \quad f_1, \ldots, f_k \in \mathrm{Vec}(S), \quad k = \dim \Delta_q,$$

where S is an open subset of M. Then the inclusion $[\bar{\Delta}, \bar{\Delta}] \subset \bar{\Delta}$ takes the form

$$[f_i, f_j](q) = \sum_{l=1}^{k} c_{ij}^l(q) f_l(q), \quad q \in S, \quad c_{ij}^l \in C^\infty(S).$$

This equality is called the *Frobenius condition*.

2.2.6 Examples

2.2.6.1 The Sub-Riemannian Problem on the Group of Motions of the Plane

The control system has the following form; see Sect. 1.1.1.4:

$$\mathcal{F} = \{u_1 f_1 + u_2 f_2 \mid (u_1, u_2) \in \mathbb{R}^2\} \subset \text{Vec}(\mathbb{R}^2 \times S^1),$$

$$f_1 = \cos\theta \frac{\partial}{\partial x} + \sin\theta \frac{\partial}{\partial y}, \qquad f_2 = \frac{\partial}{\partial\theta}.$$

The system is symmetric: $\mathcal{F} = -\mathcal{F}$. Compute its Lie algebra:

$$[f_1, f_2] = \sin\theta \frac{\partial}{\partial x} - \cos\theta \frac{\partial}{\partial y} =: f_3,$$

$$\text{Lie}_q(\mathcal{F}) = \text{span}(f_1(q), f_2(q), f_3(q)) = T_q(\mathbb{R}^2 \times S^1).$$

The system $\mathcal{F}$ is completely nonholonomic, thus controllable.

2.2.6.2 Orbits of Different Dimensions

Let

$$M = \mathbb{R}_x, \qquad \mathcal{F} = \left\{ x \frac{\partial}{\partial x} \right\} \subset \text{Vec}(M).$$

We have:

$$x_0 > 0 \quad \Rightarrow \quad O_{x_0} = \{x > 0\},$$
$$x_0 = 0 \quad \Rightarrow \quad O_{x_0} = \{x = 0\},$$
$$x_0 < 0 \quad \Rightarrow \quad O_{x_0} = \{x < 0\},$$

i.e., the system has two one-dimensional orbits and one zero-dimensional orbit.

2.2.6.3 More Orbits of Different Dimensions

Let

$$M = \mathbb{R}^3_{x,y,z}, \qquad \mathcal{F} = \left\{ x \frac{\partial}{\partial y} - y \frac{\partial}{\partial x}, \ y \frac{\partial}{\partial z} - z \frac{\partial}{\partial y}, \ z \frac{\partial}{\partial x} - x \frac{\partial}{\partial z} \right\} \subset \text{Vec}(M).$$

Then for any point $q \in \mathbb{R}^3$

$$O_q = \{(x, y, z) \in \mathbb{R}^3 \mid x^2 + y^2 + z^2 = |q|^2\},$$

this is a sphere for $q \neq 0$ and a point for $q = 0$.

An orbit of a control system is a generalisation of a trajectory of a vector field to the case of more than one vector field.

2.2.7 Exercises

1. Let $N \subset M$ be an immersed submanifold. Prove that if a vector field $f \in \text{Vec}(M)$ satisfies the condition $f(q) \in T_q N$ for all $q \in N$, then $e^{tf}(q) \in N$ for all $q \in N$, $|t| < \varepsilon$.
2. In which examples of Sect. 1.1.1 the system is small-time locally controllable?
3. Consider a control system

$$\dot{q} = f(q, u), \qquad q \in M, \quad u \in U,$$

   with a convex set $f(q_0, U)$. Prove that if $f(q_0, U) \not\ni 0$ then the system is not STLC at the point q_0.
4. Construct an example of a control system $\dot{q} = f(q, u)$, $q \in M$, $u \in U$, with a nonconvex set $f(q_0, U) \not\ni 0$ that is STLC at a point q_0.
5. Study local controllability of the system $\mathcal{F} = \{uf_1 + (1 - u)f_2 \mid u \in [0, 1]\} \subset \text{Vec}(\mathbb{R}^2)$,

$$f_1 = \frac{\partial}{\partial x}, \qquad f_2 = -\frac{\partial}{\partial x} + x^k \frac{\partial}{\partial y}, \qquad k \in \mathbb{N},$$

   at an equilibrium point.
6. Study integrability of the distribution $\Delta = \text{span}(f_1, f_2)$, $f_1 = z\frac{\partial}{\partial x} + x\frac{\partial}{\partial z}$, $f_2 = z\frac{\partial}{\partial y} + y\frac{\partial}{\partial z}$, $(x, y, z) \in \mathbb{R}^3$, $z \neq 0$. If it is integrable, describe its integral manifolds.
7. Prove that the mappings $t_i \mapsto e^{t_i f_i}(q)$ are continuous in the topology of $M^{\mathcal{F}}$; see item (7) of the proof of Theorem 2.3.
8. Fill the gaps in item (8) of the proof of Theorem 2.3.

2.3 Attainable Sets of Full-Rank Systems

Let us return to the study of attainable sets.

2.3.1   Krener's Theorem

Let $\mathcal{F} \subset \mathrm{Vec}(M)$ be a full-rank system. The assumption of full rank is not very strong in the analytic case: if it is violated, we can consider the restriction of $\mathcal{F}$ to its orbit, and this restriction is full-rank.

What is the possible structure of attainable sets of $\mathcal{F}$? It is easy to construct systems in the two-dimensional plane that have the following attainable sets:

- a smooth full-dimensional manifold without boundary; see Fig. 2.2
- a full-dimensional manifold with smooth boundary; see Fig. 2.3
- a full-dimensional manifold with non-smooth boundary, with corner or cusp singularity; see Figs. 2.4 and 2.5.

But it is impossible to construct an attainable set that is:

- a lower-dimensional submanifold; see Fig. 2.6
- a set whose boundary points are isolated from its interior points; see Fig. 2.7.

Fig. 2.2 Attainable set—smooth manifold without boundary

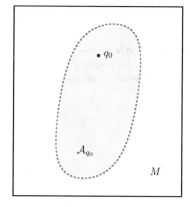

Fig. 2.3 Attainable set—manifold with smooth boundary

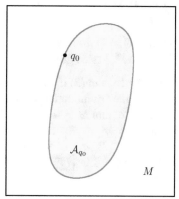

Fig. 2.4 Attainable set—manifold with nonsmooth boundary

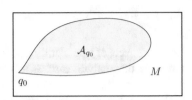

Fig. 2.5 Attainable set—manifold with nonsmooth boundary

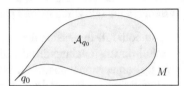

Fig. 2.6 Forbidden attainable set: subset of lower dimension

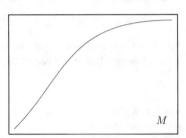

Fig. 2.7 Forbidden attainable set: subset with boundary points isolated from interior

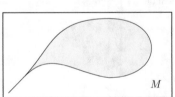

These possibilities are forbidden respectively by items (1) and (2) of the following theorem.

Theorem 2.6 (Krener) *Let $\mathcal{F} \subset \mathrm{Vec}(M)$, and let $\mathrm{Lie}_q \mathcal{F} = T_q M$ for any $q \in M$. Then:*

(1) $\mathrm{int}\, \mathcal{A}_q \neq \varnothing$ *for any* $q \in M$

(2) $\mathrm{cl}(\mathrm{int}\, \mathcal{A}_q) \supset \mathcal{A}_q$ *for any* $q \in M$.

Proof Since item (2) implies item (1), we prove item (2).

We argue by induction on dimension of M. If $\dim M = 0$, the statement is obvious. Let $\dim M > 0$.

Take any $q_1 \in \mathcal{A}_q$, and fix any neighbourhood $q_1 \in W(q_1) \subset M$. We show that $\mathrm{int}\, \mathcal{A}_q \cap W(q_1) \neq \varnothing$. There exists $f_1 \in \mathcal{F}$ such that $f_1(q_1) \neq 0$, otherwise $\mathcal{F}(q_1) = \{0\} = \mathrm{Lie}_{q_1}(\mathcal{F}) = T_{q_1} M$, a contradiction. Consider the following set for a small $\varepsilon_1 > 0$:

$$N_1 = \{e^{t_1 f_1}(q_1) \mid 0 < t_1 < \varepsilon_1\} \subset W(q_1) \cap \mathcal{A}_q.$$

N_1 is a smooth 1-dimensional manifold. If dim $M = 1$, then N_1 is open, thus $N_1 \subset$ int $\mathcal{A}_q$, so int $\mathcal{A}_q \cap W(q_1) \neq \varnothing$. Since the neighbourhood $W(q_1)$ is arbitrary, $q_1 \in$ cl(int $\mathcal{A}_q$).

Let dim $M > 1$. There exist $q_2 = e^{t_1^1 f_1}(q_1) \in N_1 \cap W(q_1)$ and $f_2 \in \mathcal{F}$ such that $f_2(q_2) \notin T_{q_2} N_1$. Otherwise dim $\mathcal{F}(q_2) = \dim \mathrm{Lie}_{q_2}(\mathcal{F}) = \dim T_{q_2} M = 1$ for any $q_2 \in N_2 \cap W$, and dim $M = 1$. Consider the following set for a small ε_2:

$$N_2 = \{e^{t_2 f_2} \circ e^{t_1 f_1}(q_2) \mid t_1^1 < t_1 < t_1^1 + \varepsilon_2,\ 0 < t_2 < \varepsilon_2\} \subset W(q_1) \cap \mathcal{A}_q.$$

N_2 is a smooth 2-dimensional manifold. If dim $M = 2$, then N_2 is open, thus $N_2 \subset$ int $\mathcal{A}_q \cap W(q_1) \neq \varnothing$ and $q_1 \in$ cl(int $\mathcal{A}_q$).

If dim $M > 2$, we proceed by induction.   □

A control system $\mathcal{F} \subset \mathrm{Vec}(M)$ is called *accessible* at a point $q \in M$ if int $\mathcal{A}_q \neq \varnothing$. It follows from Proposition 2.1 and item (1) of Krener's theorem that in the analytic case the accessibility property is equivalent to the full-rank condition.

2.3.2   Examples

Let us compute orbits and attainable sets in some problems of Sect. 1.1.1.

2.3.2.1   Stopping a Train

The control system has the form

$$\dot{x} = f_1(x) + u f_2(x), \qquad x = (x_1, x_2) \in \mathbb{R}^2, \quad |u| \leq 1,$$

$$f_1 = x_2 \frac{\partial}{\partial x_1}, \qquad f_2 = \frac{\partial}{\partial x_2}.$$

We have $[f_1, f_2] = -\frac{\partial}{\partial x_1}$, whence the system $\mathcal{F} = \{f_1 + u f_2 \mid u \in [-1, 1]\}$ is full-rank:

$$\mathrm{Lie}_x(\mathcal{F}) = \mathrm{span}\left(\frac{\partial}{\partial x_1}, \frac{\partial}{\partial x_2}\right)(x) = T_x \mathbb{R}^2 \qquad \forall x \in \mathbb{R}^2,$$

thus

$$O_x = \mathbb{R}^2 \qquad \forall x \in \mathbb{R}^2.$$

In order to find the attainable sets, we compute trajectories of the system with a constant control $u \neq 0$: they are the parabolas

$$\frac{x_2^2}{2} = ux_1 + C.$$

Now it is visually obvious that the system is controllable.

2.3.2.2 The Markov–Dubins Car

The control system has the form

$$\dot{q} = f_1(q) + uf_2(q), \qquad q = (x, y, \theta) \in M = \mathbb{R}^2 \times S^1, \quad |u| \leq 1,$$

$$f_1 = \cos\theta\frac{\partial}{\partial x} + \sin\theta\frac{\partial}{\partial y}, \qquad f_2 = \frac{\partial}{\partial \theta}.$$

We have

$$[f_1, f_2] = \sin\theta\frac{\partial}{\partial x} - \cos\theta\frac{\partial}{\partial y} =: f_3,$$

thus the system $\mathcal{F} = \{f_1 + uf_2 \mid u \in [-1, 1]\}$ is full-rank:

$$\text{Lie}_q(\mathcal{F}) = \text{span}(f_1(q), f_2(q), f_3(q)) = T_q M \qquad \forall q \in M,$$

consequently,

$$O_q = M \qquad \forall q \in M.$$

In order to describe the attainable sets, we replace the initial system $\mathcal{F}$ by a restricted system $\mathcal{F}_1 = \{f_1 \pm f_2\} \subset \mathcal{F}$ and prove that $\mathcal{F}_1$ is controllable (then $\mathcal{F}$ is controllable as well).

Trajectories of the restricted system

$$\begin{aligned}
\dot{x} &= \cos\theta, & x(0) &= x_0, \\
\dot{y} &= \sin\theta, & y(0) &= y_0, \\
\dot{\theta} &= \pm 1, & \theta(0) &= \theta_0,
\end{aligned}$$

have the form

$$\theta = \theta_0 \pm t, \qquad x = x_0 \pm (\sin(\theta_0 \pm t) - \sin\theta_0), \qquad y = y_0 \pm (\cos\theta_0 - \cos(\theta_0 \pm t)).$$

These trajectories are periodic:

$$e^{(t+2\pi n)(f_1 \pm f_2)} = e^{t(f_1 \pm f_2)}, \qquad t \in \mathbb{R}, \quad n \in \mathbb{Z}.$$

So a shift along the fields $f_1 \pm f_2$ in the negative time can be obtained as a shift in the positive time.

Consequently, if we introduce the system

$$\mathcal{F}_2 = \{f_1 \pm f_2, \ -f_1 \pm f_2\},$$

then we get

$$\mathcal{A}_q(\mathcal{F}_2) = \mathcal{A}_q(\mathcal{F}_1), \qquad q \in M.$$

But the system $\mathcal{F}_2$ is symmetric and full-rank, thus

$$\mathcal{A}_q(\mathcal{F}_2) = O_q(\mathcal{F}_2) = M,$$

whence

$$\mathcal{A}_q(\mathcal{F}) = \mathcal{A}_q(\mathcal{F}_1) = M \text{ for all } q \in M.$$

That is, the Markov–Dubins car is completely controllable in the space $\mathbb{R}^2 \times S^1$.

2.3.2.3 Euler's Elastic Problem

We have the control system

$$\dot{q} = f_1(q) + u f_2(q), \qquad q = (x, y, \theta) \in M = \mathbb{R}^2 \times S^1, \quad u \in \mathbb{R},$$

$$f_1 = \cos\theta \frac{\partial}{\partial x} + \sin\theta \frac{\partial}{\partial y}, \qquad f_2 = \frac{\partial}{\partial \theta},$$

$$q(0) = q_0 = (0, 0, 0).$$

Similarly to the Markov–Dubins car, the system $\mathcal{F} = \{f_1 + u f_2 \mid u \in \mathbb{R}\}$ satisfies

$$\text{Lie}_q(\mathcal{F}) = T_q M \Rightarrow O_q = M \qquad \forall q \in M.$$

One can easily obtain an explicit description of the attainable set from geometric arguments in the plane (x, y); see Fig. 2.8:

$$\mathcal{A}_{q_0}(t_1) = \left\{ (x, y, \theta) \in M \mid (x, y, \theta) = (t_1, 0, 0) \text{ or } x^2 + y^2 < t_1^2 \right\},$$

Fig. 2.8 Steering q_0 to q in
Euler's elastic problem

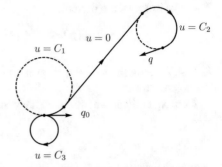

Fig. 2.9 Attainable set
$\mathcal{A}_{q_0}(t_1)$ in Euler's elastic
problem

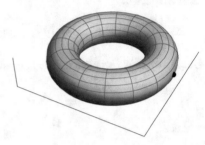

this is an open solid torus with one point at the boundary; see Fig. 2.9.

This example shows that an attainable set may be not a manifold, and neither open nor closed set.

2.3.2.4 Lie Brackets of Higher Order

Consider a control system

$$\dot{x} = 1 - u, \qquad q = (x, y) \in \mathbb{R}^2, \quad u \in [0, 1],$$
$$\dot{y} = ux^k, \qquad k \in \mathbb{N}.$$

In the vector form,

$$\dot{q} = f_1(q) + u f_2(q), \qquad q = (x, y) \in \mathbb{R}^2,$$
$$f_1 = \frac{\partial}{\partial x}, \qquad f_2 = -\frac{\partial}{\partial x} + x^k \frac{\partial}{\partial y}.$$

Let us compute the first commutator of these vector fields, linearly independent of f_1:

$$[f_1, f_2] = kx^{k-1} \frac{\partial}{\partial y},$$

$$[f_1, [f_1, f_2]] = k(k-1)x^{k-2} \frac{\partial}{\partial y},$$

$$\dots$$

$$[\underbrace{f_1, [\dots, [f_1}_{k}, f_2]\dots]] = k! \frac{\partial}{\partial y}.$$

We get

$$\mathrm{Lie}_q(\mathcal{F}) = \mathrm{span}\left(\frac{\partial}{\partial x}, \frac{\partial}{\partial y}\right)(q) = T_q \mathbb{R}^2,$$

thus $O_q = \mathbb{R}^2$ for any $q \in \mathbb{R}^2$.

It follows from geometric arguments that

- for $k = 2l + 1$ the system is controllable
- for $k = 2l$ the system has an invariant subset (e.g., the upper half-plane), thus it is not controllable.

2.3.3 Exercises

1. Construct examples of control systems having an attainable set of the following structure:

   - a smooth manifold without boundary
   - a manifold with a smooth boundary
   - a manifold with boundary having an angle singularity
   - a manifold with boundary having a cusp singularity.

Chapter 3
Optimal Control Problem

With the energy of his whole being, the boy has at last taken hold of the ox:
But how wild his will, how ungovernable his power!
At times he struts up a plateau,
When lo! he is lost again in a misty unpenetrable mountain-pass.

Pu-ming, "The Ten Oxherding Pictures" (cited by Suzuki [110])

Catching the Ox

In this chapter we study the optimal control problem. First, we present Filippov's
sufficient condition for existence of optimal controls. Then, we state the Pontryagin
maximum principle in invariant form for problems on smooth manifolds. Next,
sub-Riemannian problems are considered: we specialize the preceding results,

© The Author(s), under exclusive license to Springer Nature Switzerland AG 2022
Yu. Sachkov, *Introduction to Geometric Control*, Springer Optimization
and Its Applications 192, https://doi.org/10.1007/978-3-031-02070-4_3

discuss optimality conditions, and prove the Pontryagin maximum principle for these problems. Finally, we present a general symmetry method for construction of optimal synthesis in optimal control problems with a big symmetry group.

3.1   Optimal Control Problem: Statement and Existence of Solutions

3.1.1   Problem Statement

We consider the following *optimal control problem*:

$$\dot{q} = f(q, u), \qquad q \in M, \quad u \in U \subset \mathbb{R}^m, \tag{3.1}$$

$$q(0) = q_0, \qquad q(t_1) = q_1, \tag{3.2}$$

$$J[u] = \int_0^{t_1} \varphi(q, u)\, dt \to \min, \tag{3.3}$$

t_1 fixed or free.

A solution $q(t)$, $t \in [0, t_1]$, to this problem is said to be *(globally) optimal*.

The following assumptions are made for the dynamics $f(q, u)$:

- the mapping $q \mapsto f(q, u)$ is smooth for any $u \in U$,
- the mapping $(q, u) \mapsto f(q, u)$ is continuous for any $q \in M$, $u \in \text{cl}(U)$,
- the mapping $(q, u) \mapsto \frac{\partial f}{\partial q}(q, u)$ is continuous for any $q \in M$, $u \in \text{cl}(U)$.

The same assumptions are made for the function $\varphi(q, u)$ that determines the cost functional J.

Admissible control is $u \in L^\infty([0, t_1], U)$.

3.1.2   Reduction to the Study of Attainable Sets

In order to include the functional J into dynamics of the system, introduce a new variable equal to the running value of the cost functional along a trajectory $q_u(t)$:

$$y(t) = \int_0^t \varphi(q, u)\, dt.$$

Respectively, we introduce an extended state $\widehat{q} = \begin{pmatrix} y \\ q \end{pmatrix} \in \mathbb{R} \times M$ that satisfies an extended control system

$$\frac{d\widehat{q}}{dt} = \begin{pmatrix} \dot{y} \\ \dot{q} \end{pmatrix} = \begin{pmatrix} \varphi(q, u) \\ f(q, u) \end{pmatrix} =: \widehat{f}(\widehat{q}, u). \tag{3.4}$$

The boundary conditions for this system are

$$\widehat{q}(0) = \begin{pmatrix} 0 \\ q_0 \end{pmatrix}, \qquad \widehat{q}(t_1) = \begin{pmatrix} J \\ q_1 \end{pmatrix}.$$

Remark 3.1 A trajectory $q_{\tilde{u}}(t)$ is optimal for the problem (3.1)–(3.8) with fixed time t_1 if and only if the corresponding trajectory $\widehat{q}_{\tilde{u}}(t)$ of the extended system (3.4) comes to a point (y_1, q_1) of the attainable set $\widehat{\mathcal{A}}_{(0,q_0)}(t_1)$ such that

$$\widehat{\mathcal{A}}_{(0,q_0)}(t_1) \cap \{(y, q_1) \mid y < y_1\} = \emptyset;$$

see Fig. 3.1. For the problem with free terminal time an analogous condition is written for the attainable set $\widehat{\mathcal{A}}_{(0,q_0)}$.

Corollary 3.1 *If the attainable set $\widehat{\mathcal{A}}_{(0,q_0)}(t_1)$ is compact and $q_1 \in \mathcal{A}_{q_0}(t_1)$, then the optimal control problem (3.1)–(3.3) with fixed time t_1 has a solution.*

Theorem 3.1 (Filippov) *Suppose that control system (3.1) satisfies the hypotheses:*

(1) *the set U is compact*
(2) *the set $f(q, U)$ is convex for all $q \in M$*
(3) *there exists a compact set $K \subset M$ such that for all $q \in M \backslash K$, $u \in U$ there holds the equality $f(q, u) = 0$.*

Then the attainable sets $\mathcal{A}_{q_0}(t)$, $\mathcal{A}_{q_0}(\leq t)$ are compact for any $q_0 \in M$, $t > 0$.

Proof See [3, 46].   □

Fig. 3.1 Trajectory $q_{\tilde{u}}(t)$ is optimal

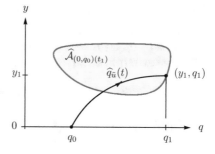

3.1.3   Existence of Optimal Controls in Optimal Control Problem

Corollary 3.2 *Let the optimal control problem* (3.1)–(3.3) *satisfy the hypotheses:*

(1) *the set U is compact*

(2) *the set $\left\{ \begin{pmatrix} \varphi(q, u) \\ f(q, u) \end{pmatrix} \mid u \in U \right\}$ is convex for all $q \in M$*

(3) *there exists a compact set $K \subset \mathbb{R} \times M$ such that $\widehat{\mathcal{A}}_{(0,q_0)}(t_1) \subset K$*

(4) *$q_1 \in \mathcal{A}_{q_0}(t_1)$.*

Then the problem (3.1)–(3.3) *with fixed time t_1 has a solution.*

Proof There exists a compact set $K' \subset \mathbb{R} \times M$ such that $K \subset$ int K'. Take a function $a \in C^\infty(\mathbb{R} \times M)$ such that

$$a|_K \equiv 1, \qquad a|_{(\mathbb{R} \times M) \setminus K'} \equiv 0.$$

Consider a new extended control system:

$$\frac{d\widehat{q}}{dt} = a(\widehat{q}) f(\widehat{q}, u), \qquad \widehat{q} \in \mathbb{R} \times M, \quad u \in U.$$

This system has compact attainable sets for time t_1 (by Theorem 3.1), which coincide with the corresponding attainable sets of extended system (3.4). Then optimal control problem (3.1)–(3.3) has a solution (by Corollary 3.1).   □

Now consider a *time-optimal problem*

$$\dot{q} = f(q, u), \qquad q \in M, \quad u \in U \subset \mathbb{R}^m, \tag{3.5}$$

$$q(0) = q_0, \qquad q(t_1) = q_1, \tag{3.6}$$

$$t_1 \to \min. \tag{3.7}$$

For this problem, Theorem 3.1 gives the following condition of existence of solution.

Corollary 3.3 *Let the following conditions hold:*

(1) *the set U is compact*

(2) *the set $f(q, U)$ is convex for all $q \in M$*

(3) *there exist $t_1 > 0$ and a compact set $K \subset M$ such that*

$$q_1 \in \mathcal{A}_{q_0}(\le t_1) \subset K.$$

Then time-optimal problem (3.5)–(3.7) *has a solution.*

3.2   The Pontryagin Maximum Principle

This section is devoted to the main necessary optimality condition for optimal control problems—the Pontryagin maximum principle (PMP).

3.2.1   Elements of Symplectic Geometry

In order to state the Pontryagin maximum principle in invariant form, we need some basic facts of symplectic geometry, which we review in this subsection.

Let M be an n-dimensional smooth manifold. Then the disjoint union of its tangent spaces

$$T M = \bigsqcup_{q \in M} T_q M = \{(q, v) \mid q \in M, \ v \in T_q M\}$$

is called its *tangent bundle*.

If $(q_1, \ldots, q_n)$ are local coordinates on M, then any tangent vector $v \in T_q M$ has a decomposition $v = \sum_{i=1}^{n} v_i \frac{\partial}{\partial q_i}$. So $(q_1, \ldots, q_n; v_1, \ldots, v_n)$ are local coordinates on $T M$, which is thus a $2n$-dimensional smooth manifold.

For any point $q \in M$, the dual space $(T_q M)^* = T_q^* M$ is called the *cotangent space* to M at q. Thus $T_q^* M$ consists of linear forms on $T_q M$. The disjoint union

$$T^* M = \bigsqcup_{q \in M} T_q^* M = \{(q, p) \mid q \in M, \ p \in T_q^* M\}$$

is called the *cotangent bundle* of M.

If $(q_1, \ldots, q_n)$ are local coordinates on M, then any covector $\lambda \in T^* M$ has a decomposition $\lambda = \sum_{i=1}^{n} p_i \, dq_i$. Thus $(q_1, \ldots, q_n; p_1, \ldots, p_n)$ are local coordinates on $T^* M$ called the *canonical coordinates*. So $T^* M$ is a smooth $2n$-dimensional manifold.

The *canonical projection* is the mapping

$$\pi : T^* M \to M, \quad T_q^* M \ni \lambda \mapsto q \in M.$$

The *Liouville (tautological) differential 1-form* $s \in \Lambda^1(T^* M)$ is defined as follows:

$$\langle s_\lambda, w \rangle = \langle \lambda, \pi_* w \rangle, \quad \lambda \in T^* M, \quad w \in T_\lambda(T^* M).$$

In the canonical coordinates on T^*M:

$$w = \sum_{i=1}^{n} \left(a_i \frac{\partial}{\partial q_i} + b_i \frac{\partial}{\partial p_i} \right),$$

$$\pi_* w = \sum_{i=1}^{n} a_i \frac{\partial}{\partial q_i},$$

$$\lambda = \sum_{i=1}^{n} p_i \, dq_i,$$

$$\langle s_\lambda, w \rangle = \sum_{i=1}^{n} p_i a_i,$$

$$s_\lambda = \sum_{i=1}^{n} p_i \, dq_i.$$

In other words, $s = p \, dq$.

The canonical *symplectic structure* on T^*M is the differential 2-form $\sigma = ds \in \Lambda^2(T^*M)$. In the canonical coordinates $\sigma = dp \wedge dq = \sum_{i=1}^{n} dp_i \wedge dq_i$.

A *Hamiltonian (Hamiltonian function)* is an arbitrary function $h \in C^\infty(T^*M)$.

The *Hamiltonian vector field* $\mathbf{h} \in \text{Vec}(T^*M)$ with the Hamiltonian function h is defined by the equality $dh = \sigma(\,\cdot\,, \mathbf{h})$. In the canonical coordinates:

$$h = h(q, p),$$

$$dh = h_q \, dq + h_p \, dp = \sum_{i=1}^{n} \left(\frac{\partial h}{\partial q_i} dq_i + \frac{\partial h}{\partial p_i} dp_i \right),$$

$$\sigma = dp \wedge dq = \sum_{i=1}^{n} dp_i \wedge dq_i,$$

$$\mathbf{h} = \frac{\partial h}{\partial p} \frac{\partial}{\partial q} - \frac{\partial h}{\partial q} \frac{\partial}{\partial p} = \sum_{i=1}^{n} \left(\frac{\partial h}{\partial p_i} \frac{\partial}{\partial q_i} - \frac{\partial h}{\partial q_i} \frac{\partial}{\partial p_i} \right).$$

The corresponding *Hamiltonian system of ODEs* is

$$\dot{\lambda} = \mathbf{h}(\lambda), \qquad \lambda \in T^*M.$$

In the canonical coordinates:

$$\begin{cases} \dot{q} = \dfrac{\partial h}{\partial p}, \\ \dot{p} = -\dfrac{\partial h}{\partial q}, \end{cases}$$

or

$$\begin{cases} \dot{q}_i = \dfrac{\partial h}{\partial p_i}, \\[2mm] \dot{p}_i = -\dfrac{\partial h}{\partial q_i}, & i = 1, \ldots, n. \end{cases}$$

The *Poisson bracket* of Hamiltonians $h, g \in C^\infty(T^*M)$ is the Hamiltonian $\{h, g\} \in C^\infty(T^*M)$ defined by the equalities

$$\{h, g\} = \mathbf{h}g = \sigma(\mathbf{h}, \mathbf{g}).$$

In the canonical coordinates:

$$\{h, g\} = \frac{\partial h}{\partial p}\frac{\partial g}{\partial q} - \frac{\partial h}{\partial q}\frac{\partial g}{\partial p} = \sum_{i=1}^n \left(\frac{\partial h}{\partial p_i}\frac{\partial g}{\partial q_i} - \frac{\partial h}{\partial q_i}\frac{\partial g}{\partial p_i} \right).$$

Notice the simplest properties of Poisson bracket.

Lemma 3.1 *Let* $a, b, c \in C^\infty(T^*M)$ *and* $\alpha, \beta \in \mathbb{R}$. *Then:*

(1) $\{a, b\} = -\{b, a\}$,
(2) $\{a, a\} = 0$,
(3) $\{\{a, b\}, c\} + \{\{b, c\}, a\} + \{\{c, a\}, b\} = 0$,
(4) $\{\alpha a + \beta b, c\} = \alpha\{a, c\} + \beta\{b, c\}$,
(5) $\{ab, c\} = \{a, c\}b + a\{b, c\}$,
(6) $[\mathbf{a}, \mathbf{b}] = \mathbf{d}, d = \{a, b\}$.

Proof

(1) $\{a, b\} = \sigma(\mathbf{a}, \mathbf{b}) = -\sigma(\mathbf{b}, \mathbf{a}) = -\{b, a\}$.
(2) $\{a, a\} = -\{a, a\} = 0$.
(3) is proved by computation in the canonical coordinates.
(4) $\{\alpha a + \beta b, c\} = (\alpha\mathbf{a} + \beta\mathbf{b})c = \alpha\{a, c\} + \beta\{b, c\}$.
(5) $\{ab, c\} = -\{c, ab\} = -\mathbf{c}(ab) = -(\mathbf{c}a)b - a(\mathbf{c}b) = -\{c, a\}b - a\{c, b\} = \{a, c\}b + a\{b, c\}$.
(6) $\mathbf{d}c = \{\{a, b\}, c\} = \{\{a, c\}, b\} + \{a, \{b, c\}\} = \mathbf{a}\mathbf{b}c - \mathbf{b}\mathbf{a}c = [\mathbf{a}, \mathbf{b}]c$, thus $\mathbf{d} = [\mathbf{a}, \mathbf{b}]$, where $d = \{a, b\}$.   □

Theorem 3.2 (Noether) *Let* $a, h \in C^\infty(T^*M)$. *Then*

$$a(e^{t\mathbf{h}}(\lambda)) \equiv \mathrm{const} \quad \Leftrightarrow \quad \{h, a\} = 0.$$

Proof $a(e^{t\mathbf{h}}(\lambda)) \equiv \mathrm{const} \Leftrightarrow \mathbf{h}a = 0 \Leftrightarrow \{h, a\} = 0$.   □

Now we describe the last construction of symplectic geometry necessary for us— *linear on fibers of* T^*M *Hamiltonians*. Let $X \in \mathrm{Vec}(M)$. The corresponding linear on fibers of T^*M Hamiltonian is defined as follows:

$$h_X(\lambda) = \langle \lambda, X(q) \rangle, \qquad q = \pi(\lambda).$$

In the canonical coordinates:

$$X = \sum_{i=1}^{n} X_i \frac{\partial}{\partial q_i},$$

$$h_X(q, p) = \sum_{i=1}^{n} p_i X_i.$$

Lemma 3.2 *Let* $X, Y \in \mathrm{Vec}(M)$. *Then:*

(1) $\{h_X, h_Y\} = h_{[X,Y]}$,
(2) $[\mathbf{h}_X, \mathbf{h}_Y] = \mathbf{h}_{[X,Y]}$,
(3) $\pi_* \mathbf{h}_X = X$.

Proof

(1) is proved by computation in the canonical coordinates.
(2) Let $g \in C^\infty(T^*M)$, then

$$[\mathbf{h}_X, \mathbf{h}_Y]g = (\mathbf{h}_X \mathbf{h}_Y - \mathbf{h}_Y \mathbf{h}_X)g = \{h_X, \{h_Y, g\}\} - \{h_Y, \{h_X, g\}\}$$

$$= \{\{h_X, h_Y\}, g\} = \mathbf{h}_{[X,Y]}g,$$

whence $[\mathbf{h}_X, \mathbf{h}_Y] = \mathbf{h}_{[X,Y]}$.
(3) In the canonical coordinates

$$h_X(q, p) = \sum_{i=1}^{n} p_i X_i(q),$$

$$\mathbf{h}_X = \sum_{i=1}^{n} \left(X_i \frac{\partial}{\partial q_i} + (\ldots) \frac{\partial}{\partial p_i} \right),$$

thus $\pi_* \mathbf{h}_X = \sum_{i=1}^{n} X_i \frac{\partial}{\partial q_i} = X$.   □

The vector field $\mathbf{h}_X \in \mathrm{Vec}(T^*M)$ is called the *Hamiltonian lift* of the vector field $X \in \mathrm{Vec}(M)$.

3.2.2 Statement of the Pontryagin Maximum Principle

Return to the optimal control problem

$$\dot{q} = f(q, u), \qquad q \in M, \quad u \in U \subset \mathbb{R}^m,$$
$$q(0) = q_0, \qquad q(t_1) = q_1,$$
$$J = \int_0^{t_1} \varphi(q, u) \, dt \to \min,$$

t_1 fixed.

Define a family of *Hamiltonians of PMP*

$$h_u^\nu(\lambda) = \langle \lambda, f(q, u) \rangle + \nu \varphi(q, u), \qquad \nu \in \mathbb{R}, \quad u \in U, \quad \lambda \in T^*M, \quad q = \pi(\lambda).$$

A fundamental necessary optimality condition for optimal control problems is given by the following statement.

Theorem 3.3 (PMP) *If a control $u(t)$ and the corresponding trajectory $q(t), t \in [0, t_1]$, are optimal, then there exist a curve $\lambda_t \in \mathrm{Lip}([0, t_1], T^*M), \lambda_t \in T^*_{q(t)}M$, and a number $\nu \leq 0$ such that the following conditions hold for almost all $t \in [0, t_1]$:*

(1) $\dot{\lambda}_t = \mathbf{h}^\nu_{u(t)}(\lambda_t),$
(2) $h^\nu_{u(t)}(\lambda_t) = \max\limits_{w \in U} h^\nu_w(\lambda_t),$
(3) $(\lambda_t, \nu) \neq (0, 0).$

Proof See [3, 30].   □

Remark 3.2 If the terminal time t_1 is free, then the following condition is added to (1)–(3):

(4) $h^\nu_{u(t)}(\lambda_t) \equiv 0.$

We will prove the Pontryagin maximum principle for sub-Riemannian problems in Sect. 3.3.6.

A curve λ_t that satisfies PMP is called an *extremal*, a curve $q(t)$—an *extremal trajectory*, a control $u(t)$—an *extremal control*.

3.2.2.1 Time-Optimal Problem

Let us apply PMP to the time-optimal problem

$$\dot{q} = f(q, u), \qquad q \in M, \quad u \in U,$$
$$q(0) = q_0, \qquad q(t_1) = q_1,$$
$$t_1 = \int_0^{t_1} 1 \, dt \to \min.$$

The Hamiltonian of PMP has the form $h_u^\nu(\lambda) = \langle \lambda, f(q, u) \rangle + \nu$. Introduce the *shortened Hamiltonian* $g_u(\lambda) = \langle \lambda, f(q, u) \rangle$. Then the statement of PMP for the time-optimal problem takes the form:

(1) $\dot{\lambda}_t = \mathbf{h}_{u(t)}^\nu(\lambda_t) = \mathbf{g}_{u(t)}(\lambda_t)$,

(2) $h_{u(t)}^\nu(\lambda_t) = \max\limits_{w \in U} h_w^\nu(\lambda_t) \quad \Leftrightarrow \quad g_{u(t)}(\lambda_t) = \max\limits_{w \in U} g_w(\lambda_t)$,

(3) $\lambda_t \neq 0$,

(4) $h_{u(t)}^\nu(\lambda_t) \equiv 0 \quad \Leftrightarrow \quad g_{u(t)}(\lambda_t) \equiv \text{const} \geq 0$.

3.2.2.2   Optimal Control Problem with General Boundary Conditions

Consider optimal control problem (3.1), (3.3), where the boundary condition (3.2) is replaced by the following more general one:

$$q(0) \in N_0, \qquad q(t_1) \in N_1. \tag{3.8}$$

Here N_0, $N_1 \subset M$ are smooth submanifolds.

For problem (3.1), (3.3), (3.8) there holds the Pontryagin maximum principle with conditions (1)–(3) of Theorem 3.3 for fixed t_1 (plus condition (4) for free t_1), with additional *transversality conditions*

(5) $\lambda_0 \perp T_{q_0} N_0, \quad \lambda_{t_1} \perp T_{q(t_1)} N_1$.

Remark 3.3 If a pair (λ_t, ν) satisfies PMP, then for any $k > 0$ the pair $(k\lambda_t, k\nu)$ also satisfies PMP.

The case $\nu < 0$ is called the *normal case*. In this case the pair (λ_t, ν) can be normalized to get $\nu = -1$.

The case $\nu = 0$ is called the *abnormal case*.

Denote the *maximized normal Hamiltonian of PMP*

$$H(\lambda) = \max\limits_{u \in U} h_u^{-1}(\lambda), \qquad \lambda \in T^*M.$$

Theorem 3.4 *Let* $H \in C^2(T^*M)$. *Then a curve* λ_t *is a normal extremal iff it is a trajectory of the Hamiltonian system* $\dot{\lambda}_t = \mathbf{H}(\lambda_t)$.

Proof See [3].   □

3.2.3   Solution to Examples of Optimal Control Problems

In this subsection we apply PMP to two optimal control problems stated in Sect. 1.1.1.

3.2.3.1 Stopping a Train

We have the time-optimal problem

$$\dot{x}_1 = x_2, \qquad x = (x_1, x_2) \in \mathbb{R}^2,$$
$$\dot{x}_2 = u, \qquad |u| \le 1,$$
$$x(0) = x^0, \qquad x(t_1) = x^1 = (0, 0),$$
$$t_1 \to \min.$$

The right-hand side of the control system $f(x, u) = (x_2, u)$ satisfies the bound

$$|f(x, u)| = \sqrt{x_2^2 + u^2} \le \sqrt{x_2^2 + 1} \le |x| + 1,$$

thus $r = x^2$ satisfies the differential inequality

$$\dot{r} = 2\langle x, \dot{x} \rangle = 2\langle x, f(x, u) \rangle \le 2(r + 1).$$

So $r(t) \le e^{2t}(r_0 + 1)$, thus attainable sets satisfy the a priori bound

$$\mathcal{A}_{x^0}(\le t) \subset \left\{ x \in \mathbb{R}^2 \mid |x| \le e^t \sqrt{(x^0)^2 + 1} \right\}.$$

Therefore we can assume that there exists a compact set $K \subset \mathbb{R}^2$ such that the right-hand side of the control system vanishes outside of K (one of conditions of the Filippov theorem).

As we showed in Sect. 2.3.2, $x^1 = (0, 0) \in \mathcal{A}_{x^0}$ for any $x^0 \in \mathbb{R}^2$.

The set of control parameters U is compact, and the set of admissible velocity vectors $f(x, U)$ is convex for any $x \in \mathbb{R}^2$. All hypotheses of the Filippov theorem are satisfied, thus optimal control exists.

We apply PMP using the canonical coordinates (p_1, p_2, x_1, x_2) on $T^*\mathbb{R}^2$. We decompose a covector $\lambda = p_1\, dx_1 + p_2\, dx_2 \in T^*\mathbb{R}^2$, then the shortened Hamiltonian of PMP reads

$$h_u(\lambda) = p_1 x_2 + p_2 u,$$

and the Hamiltonian system $\dot{\lambda} = \mathbf{h}_u(\lambda)$ reads

$$\dot{x}_1 = x_2, \qquad\qquad\qquad \dot{p}_1 = 0,$$
$$\dot{x}_2 = u, \qquad\qquad\qquad \dot{p}_2 = -p_1.$$

The maximality condition of PMP has the form

$$h_u(\lambda) = p_1 x_2 + p_2 u \to \max_{|u| \le 1},$$

and the nontriviality condition is

$$(p_1(t), p_2(t)) \neq (0, 0).$$

The Hamiltonian system implies that $p_1 \equiv$ const, thus $p_2(t)$ is linear. Moreover, $p_2(t) \not\equiv 0$ due to the nontriviality condition. Thus $p_2(t)$ has not more than one root. The maximality condition yields:

$$p_2(t) > 0 \quad \Rightarrow \quad u(t) = 1,$$
$$p_2(t) < 0 \quad \Rightarrow \quad u(t) = -1.$$

Thus extremal trajectories are the parabolas

$$x_1 = \pm \frac{x_2^2}{2} + C,$$

and the number of switchings (discontinuities) of control is not greater than 1. Let us construct such trajectories backward in time, starting from $x^1 = (0, 0)$:

- the controls $u = 1$ and $u = -1$ generate two half-parabolas terminating at x^1:

$$x_1 = \frac{x_2^2}{2}, \quad x_2 \leq 0 \quad \text{and} \quad x_1 = -\frac{x_2^2}{2}, \quad x_2 \geq 0,$$

- denote the union of these half-parabolas as Γ,
- after one switching, parabolic arcs with $u = 1$ terminating at the half-parabola $x_1 = -\frac{x_2^2}{2}$, $x_2 \geq 0$, fill the part of the plane $\mathbb{R}^2$ below the curve Γ,
- similarly, after one switching, parabolic arcs with $u = -1$ fill the part of the plane over the curve Γ.

So through each point of the plane $\mathbb{R}^2$ passes a unique extremal trajectory. In view of existence of optimal controls, the extremal trajectories are optimal.

The optimal control found has explicit dependence on the current point of the plane:

- if $x_1 = \frac{x_2^2}{2}$, $x_2 \leq 0$, or if the point (x_1, x_2) lies below the curve Γ, then $u(x_1, x_2) = 1$,
- otherwise, $u(x_1, x_2) = -1$.

The corresponding family of optimal trajectories coming to the origin is shown in Fig. 3.2.

Such a dependence $u(x)$ of optimal control on the current point x of the state space is called an *optimal synthesis*, it is the best possible form of solution to an optimal control problem.

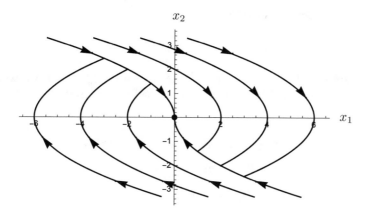

Fig. 3.2 Optimal synthesis in the problem on stopping a train

3.2.3.2 The Markov–Dubins Car

We have a time-optimal problem

$$\dot{x} = \cos\theta, \qquad q = (x, y, \theta) \in \mathbb{R}^2_{x,y} \times S^1_\theta = M,$$
$$\dot{y} = \sin\theta, \qquad |u| \le 1,$$
$$\dot{\theta} = u,$$
$$q(0) = q_0 = (0, 0, 0), \qquad q(t_1) = q_1,$$
$$t_1 \to \min.$$

The system is completely controllable, see Sect. 2.3.2.2.

All conditions of the Filippov theorem are satisfied: U is compact, $f(q, U)$ are convex, the bound $|f(q, u)| \le 2$ implies a priori bound of the attainable set. Thus optimal control exists.

We apply PMP. The vector fields

$$f_0 = \cos\theta\frac{\partial}{\partial x} + \sin\theta\frac{\partial}{\partial y},$$

$$f_1 = \frac{\partial}{\partial\theta},$$

$$f_2 = [f_0, f_1] = \sin\theta\frac{\partial}{\partial x} - \cos\theta\frac{\partial}{\partial y}$$

form a frame in $T_q M$. Define the corresponding linear on fibers of T^*M Hamiltonians:

$$h_i(\lambda) = \langle\lambda, f_i\rangle, \qquad i = 0, 1, 2.$$

The shortened Hamiltonian of PMP is

$$h_u(\lambda) = \langle \lambda, f_0 + uf_1 \rangle = h_0 + uh_1.$$

The functions h_0, h_1, h_2 form a coordinate system on T_q^*M, and we write the Hamiltonian system of PMP in the non-canonical parametrization (h_0, h_1, h_2, q) of T^*M:

$$\dot{h}_0 = \mathbf{h}_u h_0 = \{h_0 + uh_1, h_0\} = -uh_2, \tag{3.9}$$

$$\dot{h}_1 = \{h_0 + uh_1, h_1\} = h_2, \tag{3.10}$$

$$\dot{h}_2 = \{h_0 + uh_1, h_2\} = uh_0, \tag{3.11}$$

$$\dot{q} = f_0 + uf_1.$$

The maximality condition $h_u(\lambda) = h_0 + uh_1 \to \max_{|u| \leq 1}$ implies that if $h_1(\lambda_t) \neq 0$, then $u(t) = \operatorname{sgn} h_1(\lambda_t)$.

Consider the case where the control is not determined by PMP: $h_1(\lambda_t) \equiv 0$ (this case is called *singular*). Then (3.10) gives $h_2(\lambda_t) \equiv 0$, thus $h_0(\lambda_t) \neq 0$ by the nontriviality condition of PMP, so $u(t) \equiv 0$ by (3.11). The corresponding extremal trajectory $(x(t), y(t))$ is a straight line.

If $u(t) = \pm 1$, then the extremal trajectory $(x(t), y(t))$ is an arc of a unit circle. One can show [78, 98] that optimal trajectories have one of the following two types:

1. arc of unit circle + line segment + arc of unit circle
2. concatenation of three arcs of unit circles; in this case, if a, b, c are the times along the first, second, and third arc respectively, then $\pi < b < 2\pi$, $\min\{a, c\} < b$, and $\max\{a, c\} < b$.

If boundary conditions are far one from another, then the optimal trajectory has type 1 and can explicitly be constructed as follows. Draw two unit circles that satisfy the initial condition and two unit circles that satisfy the terminal condition. Draw four common tangents to the initial circles and the terminal circles, with account of direction of motion along the circles determined by the boundary conditions. Among the four constructed extremal trajectories, find the shortest one. It is the optimal trajectory.

The optimal synthesis for the Markov–Dubins car is known, but it is rather complicated, see [84].

3.2.3.3 Control of Linear Oscillator

Show that in this problem stated in Sect. 1.1.1.2 optimal trajectories are concatenations of circular arcs. Construct the optimal synthesis shown in Fig. 3.3.

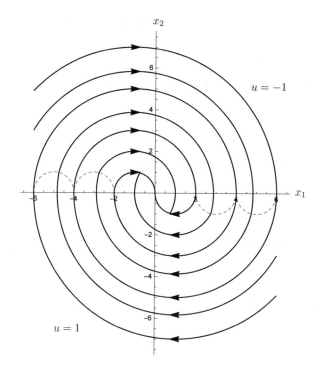

Fig. 3.3 Optimal synthesis for the controlled linear oscillator

3.3 Sub-Riemannian Geometry

In this section we consider an important special class of optimal control problems—sub-Riemannian problems.

3.3.1 Sub-Riemannian Structures and Minimizers

A *sub-Riemannian structure* on a smooth manifold M is a pair (Δ, g), where

$$\Delta = \{\Delta_q \subset T_q M \mid q \in M\}$$

is a distribution on M and

$$g = \{g_q \text{ inner product in } \Delta_q \mid q \in M\}$$

is an *inner product* (nondegenerate positive definite quadratic form) on Δ. The spaces Δ_q and inner products g_q depend smoothly on $q \in M$, and dim $\Delta_q \equiv$ const.

A curve $q \in \text{Lip}([0, t_1], M)$ is called *horizontal* (*admissible*) if

$$\dot{q}(t) \in \Delta_{q(t)} \text{ for almost all } t \in [0, t_1].$$

The *sub-Riemannian length* of a horizontal curve $q(\cdot)$ is defined as

$$l(q(\cdot)) = \int_0^{t_1} \sqrt{g(\dot{q}, \dot{q})} \, dt.$$

The *sub-Riemannian* (*Carnot–Carathéodory*) *distance* between points $q_0, q_1 \in M$ is

$$d(q_0, q_1) = \inf\{l(q(\cdot)) \mid q(\cdot) \text{ horizontal}, \ q(0) = q_0, \ q(t_1) = q_1\}.$$

A horizontal curve $q(\cdot)$ is called a *sub-Riemannian length minimizer* if

$$l(q(\cdot)) = d(q(0), q(t_1)).$$

Thus length minimizers are solutions to a *sub-Riemannian optimal control problem*:

$$\dot{q}(t) \in \Delta_{q(t)},$$
$$q(0) = q_0, \qquad q(t_1) = q_1,$$
$$l(q(\cdot)) \to \min.$$

Suppose that a sub-Riemannian structure (Δ, g) has a *global orthonormal frame* $f_1, \ldots, f_k \in \text{Vec}(M)$:

$$\Delta_q = \text{span}(f_1(q), \ldots, f_k(q)), \quad q \in M, \quad g(f_i, f_j) = \delta_{ij}, \quad i, j = 1, \ldots, k.$$

Then the optimal control problem for sub-Riemannian minimizers takes the standard form:

$$\dot{q} = \sum_{i=1}^{k} u_i f_i(q), \qquad q \in M, \quad u = (u_1, \ldots, u_k) \in \mathbb{R}^k, \tag{3.12}$$

$$q(0) = q_0, \qquad q(t_1) = q_1, \tag{3.13}$$

$$l = \int_0^{t_1} \left(\sum_{i=1}^{k} u_i^2 \right)^{1/2} dt \to \min. \tag{3.14}$$

Remark 3.4 The sub-Riemannian length does not depend on parametrization of a horizontal curve $q(t)$. Namely, if

$$\widetilde{q}(s) = q(t(s)), \qquad t(\cdot) \in \text{Lip}([0, s_1], [0, t_1]), \qquad t'(s) > 0,$$

is a reparametrization of a curve $q(t)$, then $l(\widetilde{q}(\cdot)) = l(q(\cdot))$. Indeed,

$$l(\widetilde{q}(\cdot)) = \int_0^{s_1} \sqrt{g(\widetilde{q}'(s), \widetilde{q}'(s))}\, ds = \int_0^{s_1} \sqrt{g(\dot{q}(t), \dot{q}(t))}\, t'(s)\, ds$$

$$= \int_0^{t_1} \sqrt{g(\dot{q}(t), \dot{q}(t))}\, dt = l(q(\cdot)).$$

Along with the length functional, it is convenient to consider the *energy* functional

$$J(q(\cdot)) = \frac{1}{2} \int_0^{t_1} g(\dot{q}, \dot{q})\, dt.$$

Denote $\|\dot{q}\| = \sqrt{g(\dot{q}, \dot{q})}$.

Lemma 3.3 *Let the terminal time t_1 be fixed. Then minimizers of energy are exactly length minimizers of constant velocity:*

$$J(q(\cdot)) \to \min \quad \Leftrightarrow \quad l(q(\cdot)) \to \min, \qquad \|\dot{q}\| = \text{const}.$$

Proof By the Cauchy–Schwarz inequality,

$$(l(q(\cdot)))^2 = \left(\int_0^{t_1} \|\dot{q}\| \cdot 1\, dt \right)^2 \le \int_0^{t_1} \|\dot{q}\|^2\, dt \cdot \int_0^{t_1} 1^2\, dt = 2J(q(\cdot))\, t_1,$$

moreover, equality is attained here only for $\|\dot{q}\| = \text{const}$.

It is obvious that on constant velocity curves the problems $l \to \min$ and $J \to \min$ are equivalent. And for $\|\dot{q}\| \ne \text{const}$ we have $l^2 < 2t_1 J$, i.e., J does not attain minimum. $\qquad \Box$

In the following several examples we present optimal control problems (3.12)–(3.14) for the corresponding sub-Riemannian structures.

3.3.1.1 The Sub-Riemannian Problem on the Group of Motions of the Plane

$$\dot{q} = u_1 f_1(q) + u_2 f_2(q), \qquad q = (x, y, \theta) \in \mathbb{R}^2 \times S^1, \qquad u = (u_1, u_2) \in \mathbb{R}^2,$$

$$f_1 = \cos\theta \frac{\partial}{\partial x} + \sin\theta \frac{\partial}{\partial y}, \qquad f_2 = \frac{\partial}{\partial \theta},$$

$$q(0) = q_0, \qquad q(t_1) = q_1,$$

$$l = \int_0^{t_1} \sqrt{u_1^2 + u_2^2}\, dt \to \min.$$

This problem is studied in Sect. 4.2.

3.3.1.2   The Sub-Riemannian Problem on the Heisenberg Group

$$\begin{pmatrix} \dot{x} \\ \dot{y} \\ \dot{z} \end{pmatrix} = u_1 \begin{pmatrix} 1 \\ 0 \\ -\frac{y}{2} \end{pmatrix} + u_2 \begin{pmatrix} 0 \\ 1 \\ \frac{x}{2} \end{pmatrix}, \qquad q \in \mathbb{R}^3_{x,y,z}, \quad u = (u_1, u_2) \in \mathbb{R}^2,$$

$$q(0) = q_0, \qquad q(t_1) = q_1,$$

$$l = \int_0^{t_1} \sqrt{u_1^2 + u_2^2}\, dt \to \min.$$

This problem is studied in Sect. 4.1.

3.3.1.3   The Plate-Ball Problem

$$\dot{q} = u_1 f_1(q) + u_2 f_2(q), \qquad q = (x, y, R) \in \mathbb{R}^2 \times SO(3), \quad u = (u_1, u_2) \in \mathbb{R}^2,$$

$$f_1 = \frac{\partial}{\partial x} + R \begin{pmatrix} 0 & 0 & -1 \\ 0 & 0 & 0 \\ 1 & 0 & 0 \end{pmatrix} \frac{\partial}{\partial R}, \qquad f_2 = \frac{\partial}{\partial y} + R \begin{pmatrix} 0 & 0 & 0 \\ 0 & 0 & -1 \\ 0 & 1 & 0 \end{pmatrix} \frac{\partial}{\partial R},$$

$$q(0) = q_0, \qquad q(t_1) = q_1,$$

$$l = \int_0^{t_1} \sqrt{u_1^2 + u_2^2}\, dt \to \min.$$

In the following three subsections we see how general control theory theorems specialize for sub-Riemannian (SR) problems (3.12)–(3.14).

3.3.2   The Lie Algebra Rank Condition for SR Problems

The system $\mathcal{F} = \left\{ \sum_{i=1}^k u_i f_i \mid u_i \in \mathbb{R} \right\}$ is symmetric, thus $\mathcal{A}_q = O_q$ for any $q \in M$.

Assume that M and $\mathcal{F}$ are real-analytic, and M is connected. Then by Corollary 2.3,

$$\mathcal{A}_q = M \quad \forall q \in M \Leftrightarrow O_q = M \quad \forall q \in M$$

$$\Leftrightarrow \operatorname{Lie}_q(\mathcal{F}) = \operatorname{Lie}_q(f_1, \ldots, f_k) = T_q M \quad \forall\, q \in M.$$

3.3.3 The Filippov Theorem for SR Problems

We can equivalently rewrite the optimal control problem (3.12)–(3.14) for SR minimizers as the following time-optimal problem:

$$\dot{q} = \sum_{i=1}^{k} u_i f_i(q), \qquad \sum_{i=1}^{k} u_i^2 \leq 1, \quad q \in M,$$

$$q(0) = q_0, \qquad q(t_1) = q_1,$$

$$t_1 \to \min.$$

Let us check hypotheses of the Filippov theorem (Corollary 3.3) for this problem. The set of control parameters $U = \{u \in \mathbb{R}^k \mid \sum_{i=1}^{k} u_i^2 \leq 1\}$ is compact, and the sets of admissible velocities $\left\{ \sum_{i=1}^{k} u_i f_i(q) \mid u \in U \right\} \subset T_q M$ are convex. If we prove an a priori estimate for the attainable sets $\mathcal{A}_{q_0}(\leq t_1)$, then the Filippov theorem guarantees existence of length minimizers.

3.3.4 The Pontryagin Maximum Principle for SR Problems

Introduce the linear on fibers of T^*M Hamiltonians $h_i(\lambda) = \langle \lambda, f_i \rangle$, $i = 1, \ldots, k$. Then the Hamiltonian of PMP for SR problem takes the form

$$h_u^\nu(\lambda) = \sum_{i=1}^{k} u_i h_i(\lambda) + \frac{\nu}{2} \sum_{i=1}^{k} u_i^2.$$

3.3.4.1 The Normal Case

Let $\nu = -1$. The maximality condition

$$\sum_{i=1}^{k} u_i h_i - \frac{1}{2} \sum_{i=1}^{k} u_i^2 \to \max_{u_i \in \mathbb{R}}$$

yields $u_i = h_i$, then the Hamiltonian takes the form

$$h_u^{-1}(\lambda) = \frac{1}{2} \sum_{i=1}^{k} h_i^2(\lambda) =: H(\lambda).$$

The function $H(\lambda)$ is called the *normal maximized Hamiltonian*. Since it is smooth, in the normal case extremals satisfy the Hamiltonian system $\dot\lambda = \mathbf{H}(\lambda)$.

3.3.4.2 The Abnormal Case

Let $\nu = 0$. The maximality condition

$$\sum_{i=1}^{k} u_i h_i \to \max_{u_i \in \mathbb{R}}$$

implies that $h_i(\lambda_t) \equiv 0$, $\quad i = 1, \dots, k$. Thus abnormal extremals satisfy the conditions:

$$\dot\lambda_t = \sum_{i=1}^{k} u_i(t) \mathbf{h}_i(\lambda_t),$$

$$h_1(\lambda_t) = \cdots = h_k(\lambda_t) \equiv 0.$$

Remark 3.5 Normal length minimizers are projections of solutions to the smooth Hamiltonian system $\dot\lambda = \mathbf{H}(\lambda)$, thus they are smooth. An important open question of sub-Riemannian geometry is whether abnormal length minimizers are smooth, see [1, 48].

We prove PMP for SR problems in Sect. 3.3.6.

3.3.5 Optimality of SR Extremal Trajectories

In this subsection we consider normal extremal trajectories $q(t) = \pi(\lambda_t)$, $\dot\lambda_t = \mathbf{H}(\lambda_t)$.

A horizontal curve $q(t)$ is called a *SR geodesic* if $g(\dot q, \dot q) \equiv$ const and short arcs of $q(t)$ are optimal.

Theorem 3.5 (Legendre) *Normal extremal trajectories are SR geodesics.*

Proof See [2].   □

Example 3.1 Geodesics on S^2: consider the standard sphere $S^2 \subset \mathbb{R}^3$ with the Riemannian metric induced by the Euclidean metric of $\mathbb{R}^3$. Geodesics starting from the North pole $N \in S^2$ are great circles at the sphere passing through N (meridians). Such geodesics are optimal up to the South pole $S \in S^2$. Variation of geodesics passing through N yields the fixed point S, thus S is a conjugate point to N. On the other hand, S is the intersection point of different geodesics of the same length starting at N, thus S is a Maxwell point. In this example, a conjugate point coincides with a Maxwell point due to the one-parameter group of symmetries (rotations of S^2 around the line $NS \subset \mathbb{R}^3$). In order to distinguish these points, one should destroy the rotational symmetry as in the following example.

Example 3.2 Geodesics on an ellipsoid: consider a three-axes ellipsoid with the Riemannian metric induced by the Euclidean metric of the ambient $\mathbb{R}^3$. Construct the family of geodesics on the ellipsoid starting from a vertex N, and let us look at this family from the opposite vertex S. The family of geodesics has an envelope—an astroid centred at S. Each point of the astroid is a conjugate point. At such points the geodesics lose their local optimality. On the other hand, there is a segment joining a pair of opposite vertices of the astroid, where pairs of geodesics of the same length meet one another. This segment (except its endpoints) consists of Maxwell points. At such points geodesics on the ellipsoid lose their global optimality.

We will clarify now the notions and facts that appear in these examples.

Consider the normal Hamiltonian system of PMP $\dot{\lambda}_t = \mathbf{H}(\lambda_t)$. The Hamiltonian H is an integral of this system. We can assume that $H(\lambda_t) \equiv \frac{1}{2}$, this corresponds to the *arclength parametrization* of normal geodesics: $\|\dot{q}(t)\| \equiv 1$. Denote the cylinder $C = T_{q0}^* M \cap \{H = \frac{1}{2}\}$ and define the sub-Riemannian *exponential mapping*

$$\text{Exp} : C \times \mathbb{R}_+ \to M,$$

$$\text{Exp}(\lambda_0, t) = \pi \circ e^{t\mathbf{H}}(\lambda_0) = q(t).$$

A point $\text{Exp}(\lambda_0, t_1)$ is called a *conjugate point* along the geodesic $q(t) = \text{Exp}(\lambda_0, t)$ if it is a critical value of Exp, i.e., $\text{Exp}_{*(\lambda_0, t_1)}$ is degenerate. A point $\text{Exp}(\lambda_0, t_1)$ is conjugate iff the Jacobian of the exponential mapping vanishes:

$$\det\left(\frac{\partial \text{Exp}}{\partial(\lambda_0, t)}\right)\bigg|_{t=t_1} = 0.$$

At a conjugate point a geodesic is tangent to the envelope of the family of geodesics starting from the initial point q_0.

A trajectory $q(t)$ of control system (3.1) with a control $u(t)$ and boundary conditions (3.2) is called *locally (strongly) optimal* if there is $\varepsilon > 0$ such that

$$J[u] \leq J[\tilde{u}]$$

for any admissible control $\tilde{u}(t)$ such that the corresponding trajectory $\tilde{q}(t) = q_{\tilde{u}}(t)$ satisfies boundary conditions (3.2) and the inequality

$$\max_{t \in [0, t_1]} |q(t) - \tilde{q}(t)| < \varepsilon$$

in local coordinates on M.

Theorem 3.6 (Jacobi) *Let a normal geodesic $q(t)$ be a projection of a unique, up to a scalar multiple, extremal. Then $q(t)$ loses its local optimality at the first conjugate point.*

Proof See [2].   □

A point $q(t)$ is called a *Maxwell point* along a geodesic $q(s) = \mathrm{Exp}(\lambda_0, s)$ if there exists another geodesic $\tilde{q}(s) = \mathrm{Exp}(\widetilde{\lambda_0}, s) \neq q(s)$ such that $q(t) = \tilde{q}(t)$. See Fig. 3.4: there exists a geodesic $\widehat{q}(s)$ coming to the point $q_1 = q(t_1)$ earlier than $q(s)$.

Lemma 3.4 *If M and H are real-analytic, then a normal geodesic cannot be optimal after a Maxwell point.*

Proof Let $q_1 = q(t_1)$ be a Maxwell point along a geodesic $q(t) = \mathrm{Exp}(\lambda_0, t)$, and let $\tilde{q}(t) = \mathrm{Exp}(\tilde{\lambda}_0, t) \neq q(t)$ be another geodesic with $\tilde{q}(t_1) = q_1$. If $q(t)$, $t \in [0, t_1 + \varepsilon]$, $\varepsilon > 0$, is optimal, then the following curve is optimal as well:

$$\bar{q}(t) = \begin{cases} \tilde{q}(t), & t \in [0, t_1], \\ q(t), & t \in [t_1, t_1 + \varepsilon]. \end{cases}$$

The geodesics $q(t)$ and $\bar{q}(t)$ coincide at the segment $t \in [t_1, t_1 + \varepsilon]$. Since they are analytic, they should coincide at the whole domain $t \in [0, t_1 + \varepsilon]$. Thus $q(t) \equiv \tilde{q}(t)$, $t \in [0, t_1]$, a contradiction.   □

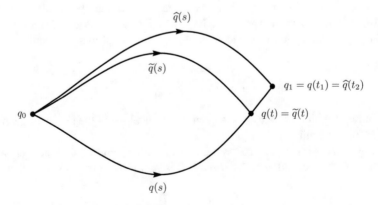

Fig. 3.4 Maxwell point $q(t)$: $t_2 < t_1$

Theorem 3.7 *Let $q(t)$ be a normal geodesic that is a projection of a unique, up to a scalar multiple, extremal. Then $q(t)$ loses its global optimality either at the first Maxwell point or at the first conjugate point (at the first one of these two points).*

Proof See [2]. □

3.3.6 Proof of the Pontryagin Maximum Principle for Sub-Riemannian Problems

In this subsection we prove PMP for the sub-Riemannian optimal control problem (3.12)–(3.14):

$$\dot{q} = \sum_{i=1}^{k} u_i f_i(q) =: f_u(q), \qquad q \in M, \quad u = (u_1, \ldots, u_k) \in \mathbb{R}^k,$$

$$q(0) = q_0, \qquad q(t_1) = q_1,$$

$$l = \int_0^{t_1} \left(\sum_{i=1}^{k} u_i^2 \right)^{1/2} dt \to \min.$$

As we showed in Sect. 3.3.4, the Pontryagin maximum principle (Theorem 3.3) for this problem takes the following form.

Theorem 3.8 (PMP for SR Problems) *Let $\overline{q} \in \mathrm{Lip}([0, t_1], M)$ be a SR minimizer for which the corresponding control $\overline{u}(t)$ satisfies the condition $\sum_{i=1}^{k} \overline{u}_i^2(t) \equiv$ const. Then there exists a curve $\lambda_t \in \mathrm{Lip}([0, t_1], T^*M), \pi(\lambda_t) = \overline{q}(t)$, such that for almost all $t \in [0, t_1]$*

$$\dot{\lambda}_t = \sum_{i=1}^{k} \overline{u}_i(t) \mathbf{h}_i(\lambda_t), \tag{3.15}$$

and one of the conditions hold:

(N) $h_i(\lambda_t) \equiv \overline{u}_i(t), \quad i = 1, \ldots, k,$ *or*
(A) $h_i(\lambda_t) \equiv 0, \quad i = 1, \ldots, k, \qquad \lambda_t \neq 0 \qquad \forall t \in [0, t_1].$

In conditions (N), (A) corresponding to the normal and abnormal cases, as always, $h_i(\lambda) = \langle \lambda, f_i \rangle, i = 1, \ldots, k$.

Theorem 3.8 follows from the next two theorems.

Theorem 3.9 *Let the hypotheses of Theorem* 3.8 *hold. For any* $t \in [0, t_1]$, *let* P_t : $M \to M$ *denote the flow of the nonautonomous vector field* $f_{\bar{u}(t)} = \sum_{i=1}^{k} \bar{u}_i(t) f_i$ *from the time* 0 *to the time* t. *Then there exists* $\lambda_0 \in T_{q_0}^* M$ *such that the curve*

$$\lambda_t = (P_t^{-1})^*(\lambda_0) \in T_{\bar{q}(t)}^* M \tag{3.16}$$

satisfies one of conditions (N), (A) *of Theorem* 3.8.

Theorem 3.10 *Let the hypotheses of Theorems* 3.8 *and* 3.9 *hold. Then ODE* (3.15) *follows from identity* (3.16).

It is obvious that Theorem 3.8 follows from Theorems 3.9 and 3.10.

Remark 3.6 In Theorem 3.9, the flow P_t : $M \to M$ of the nonautonomous field $f_{\bar{u}(t)}$ from the time 0 to the time t is given as follows:

$$P_t(q) = \bar{q}(t), \qquad q \in M, \quad t \in [0, t_1],$$

$$\frac{d}{dt}\bar{q}(t) = \sum_{i=1}^{k} \bar{u}_i(t) f_i(\bar{q}(t)), \qquad \bar{q}(0) = q.$$

Further, in Theorem 3.9 we use the mapping $(P_t^{-1})^*$: $T_{q_0}^* M \to T_{\bar{q}(t)}^* M$, recall the necessary definition. If $F : M \to N$ is a smooth mapping between smooth manifolds and $q \in M$, then there is defined the differential

$$F_{*q} : T_q M \to T_{F(q)} N,$$

and the dual mapping of cotangent spaces:

$$F_q^* = (F_{*q})^* : T_{F(q)}^* N \to T_q^* M,$$

$$\langle F_q^*(\lambda), v \rangle = \langle \lambda, F_{q*}(v) \rangle, \qquad v \in T_q M, \quad \lambda \in T_{F(q)}^* N.$$

Now we prove Theorem 3.9.

Proof The curve $\bar{q}(t)$ is a minimizer of the length functional l of constant velocity, thus it is a minimizer of the energy functional $J(u) = \frac{1}{2} \int_0^{t_1} \sum_{i=1}^{k} u_i^2(t) \, dt$ for a fixed t_1.

Take any control $u(\cdot) = \bar{u}(\cdot) + v(\cdot) \in L^\infty([0, t_1], \mathbb{R}^k)$ and consider the corresponding Cauchy problem

$$\dot{q}(t) = f_{u(t)}(q(t)), \qquad q(0) = q_0.$$

Recall that P_t : $M \to M$ is the flow of the nonautonomous vector field $f_{\bar{u}(t)}$ from the time 0 to the time t. Consider the curve $x(t) = P_t^{-1}(q(t))$ and derive an ODE for $x(t)$. We differentiate the identity $q(t) = P_t(x(t))$ and get

$$\dot{q}(t) = f_{\bar{u}(t)}(P_t(x(t))) + (P_t)_*\dot{x}(t),$$

whence

$$\begin{aligned}
\dot{x}(t) &= (P_t^{-1})_*[\dot{q}(t) - f_{\bar{u}(t)}(P_t(x(t)))] \\
&= (P_t^{-1})_*[(f_{u(t)} - f_{\bar{u}(t)})(P_t(x(t)))] \\
&= [(P_t^{-1})_*(f_{u(t)-\bar{u}(t)})](x(t)) \\
&= [(P_t^{-1})_* f_{v(t)}](x(t)).
\end{aligned}$$

We denote $g_v^t = (P_t^{-1})_* f_v$ and get the required ODE

$$\dot{x}(t) = g_{v(t)}^t(x(t)), \qquad x(0) = P_0^{-1}(q_0) = q_0. \tag{3.17}$$

Notice that f_v is linear in v, thus g_v^t is linear in v.

For any $v \in L^\infty([0, t_1], \mathbb{R}^k)$, consider a mapping

$$\mathbb{R} \ni s \mapsto \begin{pmatrix} x(t_1; \bar{u} + sv) \\ J(\bar{u} + sv) \end{pmatrix} \in M \times \mathbb{R},$$

where $x(t_1; \bar{u} + sv)$ is the solution to Cauchy problem (3.17) corresponding to the control $\bar{u} + sv$, and $J(\bar{u} + sv)$ is the corresponding energy.

Lemma 3.5 *There exists a covector $\bar{\lambda} \in (T_{q_0}M \oplus \mathbb{R})^*$, $\bar{\lambda} \neq 0$, such that for any $v \in L^\infty([0, t_1], \mathbb{R}^k)$ there holds the equality*

$$\left\langle \bar{\lambda}, \left(\frac{\partial x(t_1; \bar{u} + sv)}{\partial s}\bigg|_{s=0}, \frac{\partial J(\bar{u} + sv)}{\partial s}\bigg|_{s=0} \right) \right\rangle = 0. \tag{3.18}$$

Proof Denote

$$\Phi(v) = \left(\frac{\partial x(t_1; \bar{u} + sv)}{\partial s}\bigg|_{s=0}, \frac{\partial J(\bar{u} + sv)}{\partial s}\bigg|_{s=0} \right),$$

$$\Phi : L^\infty([0, t_1], \mathbb{R}^k) \to T_{q_0}M \oplus \mathbb{R}.$$

We compute the derivatives in the definition of the mapping Φ. It is easy to see that

$$\frac{\partial J(\bar{u} + sv)}{\partial s}\bigg|_{s=0} = \int_0^{t_1} \sum_{i=1}^k \bar{u}_i(t)v_i(t)\,dt. \tag{3.19}$$

Indeed, this follows from the expansion

$$J(\overline{u} + sv) = \frac{1}{2} \int_0^{t_1} |\overline{u} + sv|^2 \, dt$$

$$= \frac{1}{2} \int_0^{t_1} \left(|\overline{u}|^2 + 2s \sum_{i=1}^k \overline{u}_i(t) v_i(t) + s^2 |v|^2 \right) dt.$$

Further, we show that

$$\left. \frac{\partial x(t_1; \overline{u} + sv)}{\partial s} \right|_{s=0} = \int_0^{t_1} g_{v(t)}^t(q_0) \, dt = \int_0^{t_1} \sum_{i=1}^k ((P_t^{-1})_* f_i)(q_0) v_i(t) \, dt.$$

$$(3.20)$$

The ODE $\dot{x}(t; \overline{u} + sv) = g_{sv}^t(x(t; \overline{u} + sv))$ implies in local coordinates that

$$x(t_1; \overline{u} + sv) = q_0 + \int_0^{t_1} g_{sv(t)}^t(x(t; \overline{u} + sv)) \, dt$$

$$= q_0 + s \int_0^{t_1} g_{v(t)}^t(x(t; \overline{u} + sv)) \, dt,$$

whence

$$\left. \frac{\partial x(t_1; \overline{u} + sv)}{\partial s} \right|_{s=0} = \int_0^{t_1} g_{v(t)}^t(x(t; \overline{u})) \, dt$$

$$= \int_0^{t_1} g_{v(t)}^t(q_0) \, dt = \int_0^{t_1} \sum_{i=1}^k ((P_t^{-1})_* f_i)(q_0) v_i(t) \, dt.$$

One can see from (3.19), (3.20) that the mapping Φ is linear. We show that it is not surjective. By contradiction, let $\text{Im}\,\Phi = T_{q_0} M \oplus \mathbb{R}$, then there exist $v^0, \ldots, v^n \in L^\infty([0, t_1], \mathbb{R}^k)$ such that $\Phi(v^0), \ldots, \Phi(v^n)$ are linearly independent, i.e., the vectors

$$\left(\begin{array}{c} \left. \frac{\partial x(t_1; \overline{u} + sv^0)}{\partial s} \right|_{s=0} \\ \left. \frac{\partial J(\overline{u} + sv^0)}{\partial s} \right|_{s=0} \end{array} \right), \quad \ldots, \quad \left(\begin{array}{c} \left. \frac{\partial x(t_1; \overline{u} + sv^n)}{\partial s} \right|_{s=0} \\ \left. \frac{\partial J(\overline{u} + sv^n)}{\partial s} \right|_{s=0} \end{array} \right)$$

are linearly independent. Consider the mapping

$$F \; : \; (s_0, \ldots, s_n) \mapsto \left(\begin{array}{c} x\left(t_1; \overline{u} + \sum_{i=0}^n s_i v^i \right) \\ J\left(\overline{u} + \sum_{i=0}^n s_i v^i \right) \end{array} \right), \qquad \mathbb{R}^{n+1} \to M \times \mathbb{R}.$$

The mapping F is smooth near the point $0 \in \mathbb{R}^{n+1}$ and has a nondegenerate Jacobian at this point. Thus there exists a neighbourhood $O_0 \subset \mathbb{R}^{n+1}$ such that the restriction $F|_{O_0}$ is a diffeomorphism.

Consequently,

$$F(0) = \begin{pmatrix} x(t_1; \overline{u}) \\ J(\overline{u}) \end{pmatrix} = \begin{pmatrix} q_0 \\ J(\overline{u}) \end{pmatrix} \in \mathrm{int}\ F(O_0).$$

Thus there exists a control $v(\cdot) = \sum_{i=0}^{n} s_i v^i(\cdot)$ for which

$$x(t_1; \overline{u} + v) = q_0, \qquad J(\overline{u} + v) < J(\overline{u}).$$

Consider the corresponding trajectory $t \mapsto q(t; \overline{u} + v)$. We have

$$q(0; \overline{u} + v) = q_0,$$

$$q(t_1; \overline{u} + v) = P_{t_1}(x(t_1; \overline{u} + v)) = P_{t_1}(q_0) = q_1.$$

So the curve $q(t; \overline{u} + v)$ connects the points q_0 and q_1 with a lesser value of the functional J than the optimal trajectory $\overline{q}(t) = q(t; \overline{u})$. The contradiction obtained completes the proof of this lemma. $\qquad\square$

We continue the proof of Theorem 3.9. By the previous lemma, there exists a covector $0 \neq \overline{\lambda} \in (T_{q_0}M \oplus \mathbb{R})^*$ such that for any $v \in L^\infty([0, t_1], \mathbb{R}^k)$ we have

$$\left\langle \overline{\lambda}, \left(\left. \frac{\partial x(t_1; \overline{u} + sv)}{\partial s} \right|_{s=0}, \left. \frac{\partial J(\overline{u} + sv)}{\partial s} \right|_{s=0} \right) \right\rangle = 0.$$

It is obvious that if this condition holds for some covector $\overline{\lambda}$, then it also holds for any covector $\alpha\overline{\lambda}$, $\alpha \neq 0$. Consequently, we can choose a covector $\overline{\lambda}$ of the form

$$\overline{\lambda} = (\lambda_0, -1) \qquad \text{or} \qquad \overline{\lambda} = (\lambda_0, 0), \quad \lambda_0 \neq 0.$$

Thus there exists a covector $\lambda_0 \in T_{q_0}^*M$ such that for any $v \in L^\infty([0, t_1], \mathbb{R}^k)$

$$\left. \frac{\partial J(\overline{u} + sv)}{\partial s} \right|_{s=0} - \left\langle \lambda_0, \left. \frac{\partial x(t_1; \overline{u} + sv)}{\partial s} \right|_{s=0} \right\rangle = 0 \qquad (3.21)$$

or

$$0 = \left\langle \lambda_0, \left. \frac{\partial x(t_1; \overline{u} + sv)}{\partial s} \right|_{s=0} \right\rangle, \qquad \lambda_0 \neq 0. \qquad (3.22)$$

Consider the case (3.21). Equalities (3.19) and (3.20) imply that for any $v \in L^{\infty}([0, t_1], \mathbb{R}^k)$

$$\int_0^{t_1} \sum_{i=1}^{k} \bar{u}_i(t) v_i(t)\, dt = \int_0^{t_1} \sum_{i=1}^{k} \left\langle \lambda_0, ((P_t^{-1})_* f_i)(q_0) \right\rangle v_i(t)\, dt$$

$$= \int_0^{t_1} \sum_{i=1}^{k} \langle \lambda_t, f_i(\bar{q}(t)) \rangle v_i(t)\, dt$$

$$= \int_0^{t_1} \sum_{i=1}^{k} h_i(\lambda_t) v_i(t)\, dt.$$

Since the functions $v_i \in L^{\infty}[0, t_1]$ are arbitrary, we get in case (3.21)

(N) $\bar{u}_i(t) = h_i(\lambda_t),$ $i = 1, \ldots, k.$

Similarly, in case (3.22) we get the condition

(A) $0 = h_i(\lambda_t),$ $i = 1, \ldots, k;$ $\lambda_0 \neq 0.$

Theorem 3.9 is proved. $\square$

Now we prove Theorem 3.10.

Proof Recall: we should show that the curve $\lambda_t = (P_t^{-1})^* \lambda_0 \in T_{\bar{q}(t)}^* M$ satisfies the ODE $\dot{\lambda}_t = \sum_{i=1}^{k} \bar{u}_i(t) \mathbf{h}_i(\lambda_t)$. Now we prove this for the flow of an autonomous vector field.

Lemma 3.6 *Let* $X \in \mathrm{Vec}(M)$, $P_t = e^{tX}$. *Then the curve* $\lambda_t = (P_t^{-1})^* \lambda_0$ *satisfies the ODE* $\dot{\lambda}_t = \mathbf{h}_X(\lambda_t)$. $\square$

Proof We set $\varphi(t) = (P_t^{-1})^* (\lambda_0)$, then we have to prove that

$$\dot{\varphi}(t) = \mathbf{h}_X(\varphi(t)) \in T_{\varphi(t)}(T^* M).$$

A function $a \in C^{\infty}(T^* M)$ is called *constant on fibers of* $T^* M$ if it has the form $a = \alpha \circ \pi$ for some function $\alpha \in C^{\infty}(M)$. Notation: $a \in C_{\mathrm{cst}}^{\infty}(T^* M)$.

A function $h_Y \in C^{\infty}(T^* M)$ is called *linear on fibers of* $T^* M$ if

$$h_Y(\lambda) = \langle \lambda, Y(q) \rangle, \qquad q = \pi(\lambda), \quad \lambda \in T^* M,$$

for some vector field $Y \in \mathrm{Vec}(M)$. Notation: $h_Y \in C_{\mathrm{lin}}^{\infty}(T^* M)$.

An *affine on fibers of* $T^* M$ *function* is a sum of a constant on fibers and a linear on fibers functions:

$$C_{\mathrm{aff}}^{\infty}(T^* M) = C_{\mathrm{cst}}^{\infty}(T^* M) + C_{\mathrm{lin}}^{\infty}(T^* M).$$

Remark 3.7 Let $v, \omega \in T_\lambda(T^*M)$. The equality $v = \omega$ holds if and only if

$$vg = \omega g \qquad \forall g \in C_{\mathrm{aff}}^\infty(T^*M).$$

Indeed, the value $vg = \langle d_\lambda g, v \rangle$ depends only on the first order Taylor polynomial of the function g.   □

So we check the required equality $\dot{\varphi}(t) = \mathbf{h}_X(\varphi(t))$ for affine on fibers of T^*M functions.

Let $a = \alpha \circ \pi \in C_{\mathrm{cst}}^\infty(T^*M)$, we check the equality $\dot{\varphi}(t)a = \mathbf{h}_X a$. We have

$$\mathbf{h}_X a = \{h_X, a\} = \sum_{i=1}^n \frac{\partial h_X}{\partial p_i} \frac{\partial \alpha}{\partial q_i} = \sum_{i=1}^n X_i \frac{\partial \alpha}{\partial q_i} = X\alpha,$$

$$\dot{\varphi}(t)a = \frac{d}{dt} a(\varphi(t)) = \frac{d}{dt} \alpha \circ e^{tX}(q_0) = (X\alpha)(\varphi(t)),$$

and the required equality is proved for functions $a \in C_{\mathrm{cst}}^\infty(T^*M)$.

Now let $h_Y \in C_{\mathrm{lin}}^\infty(T^*M)$, we check the equality $\dot{\varphi}(t)h_Y = \mathbf{h}_X h_Y$. We have

$$\mathbf{h}_X h_Y = \{h_X, h_Y\} = h_{[X,Y]}.$$

On the other hand,

$$
\begin{aligned}
\dot{\varphi}(t)h_Y &= \frac{d}{dt} h_Y \circ \varphi(t) = \left.\frac{d}{d\tau}\right|_{\tau=0} h_Y \circ \varphi(t+\tau) \\
&= \left.\frac{d}{d\tau}\right|_{\tau=0} h_Y \circ (e^{-\tau X})^* \circ (e^{-tX})^*(\lambda_0) \\
&= \left.\frac{d}{d\tau}\right|_{\tau=0} \left\langle (e^{-\tau X})^* \circ (e^{-tX})^*(\lambda_0), Y(e^{(t+\tau)X}(q_0)) \right\rangle \\
&= \left\langle \varphi(t), \left.\frac{d}{d\tau}\right|_{\tau=0} e_*^{-\tau X} Y(e^{\tau X} \circ e^{tX}(q_0)) \right\rangle \\
&= \left\langle \varphi(t), [X, Y](e^{tX}(q_0)) \right\rangle = h_{[X,Y]}(\varphi(t)).
\end{aligned}
$$

In the penultimate transition we used the equality

$$\left.\frac{d}{d\tau}\right|_{\tau=0} e_*^{-\tau X} Y(e^{\tau X}(q)) = [X, Y](q), \qquad (3.23)$$

which we prove now.

We have

$$\frac{d}{d\tau}\Big|_{\tau=0} e_*^{-\tau X} Y(e^{\tau X}(q)) = \frac{\partial^2}{\partial\tau\partial s}\Big|_{\tau=0,s=0} e^{-\tau X} \circ e^{sY} \circ e^{\tau X}(q).$$

We compute Taylor expansions of the compositions in the right-hand side:

$$e^{\tau X}(q) = q + \tau X(q) + o(\tau),$$

$$e^{sY} \circ e^{\tau X} = e^{sY}(q + \tau X(q) + o(\tau))$$

$$= q + \tau X(q) + o(\tau) + sY(q + \tau X(q) + o(\tau)) + o(s)$$

$$= q + \tau X(q) + sY(q) + s\tau\frac{\partial Y}{\partial q}X(q) + \dots,$$

$$e^{-\tau X} \circ e^{sY} \circ e^{\tau X}(q) = q + \tau X(q) + sY(q) + s\tau\frac{\partial Y}{\partial q}X(q)$$

$$- \tau X(q) - \tau s\frac{\partial X}{\partial q}Y(q) + \dots$$

$$= q + sY(q) + s\tau[X, Y](q) + \dots,$$

thus

$$\frac{\partial^2}{\partial\tau\partial s}\Big|_{\tau=0,s=0} e^{-\tau X} \circ e^{sY} \circ e^{\tau X}(q) = [X, Y](q),$$

and equality (3.23) follows.

Lemma 3.6 is proved.   □

Similarly to Lemma 3.6 for an autonomous vector field X, one proves the equality $\dot{\lambda}_t = \sum_{i=1}^{k} \overline{u}_i(t)\mathbf{h}_i(\lambda_t)$ for a curve $\lambda_t = (P_t^{-1})^*\lambda_0$ in the case of a nonautonomous vector field $f_{\overline{u}(t)}$.

This completes the proof of Theorem 3.10.   □

As we noticed above, Theorem 3.8 follows from Theorems 3.9 and 3.10.

The Pontryagin maximum principle for SR problems is proved.

3.4   A Symmetry Method for Construction of Optimal Synthesis

We describe a general method for construction of optimal synthesis for sub-Riemannian problems with a big group of symmetries (e.g. for left-invariant SR problems on Lie groups). This method easily generalizes from SR problems to more general classes of optimal control problems.

Assume that for any $q_1 \in M$ there exists a length minimizer $q(t)$ that connects q_0 and q_1. Moreover, suppose for simplicity that all abnormal geodesics are simultaneously normal. Thus all geodesics are parametrised by the normal exponential mapping

$$\text{Exp} : N \to M, \qquad N = C \times \mathbb{R}_+, \qquad C = T^*_{q_0} M \cap \left\{ H = \frac{1}{2} \right\}.$$

If this mapping is bijective onto $M \setminus \{q_0\}$, then any point $q_1 \in M$ is connected with q_0 by a unique geodesic $q(t)$, and by virtue of existence of length minimizers this geodesic is optimal.

But typically the exponential mapping is not bijective due to Maxwell points. Denote by $t^1_{\text{Max}}(\lambda) \in (0, +\infty]$ the first Maxwell time along a geodesic $\text{Exp}(\lambda, t)$, $\lambda \in C$. Consider the Maxwell set in the image of the exponential mapping

$$\text{Max} = \left\{ \text{Exp}(\lambda, t^1_{\text{Max}}(\lambda)) \mid \lambda \in C \right\},$$

and introduce the restricted exponential mapping

$$\text{Exp} : \widetilde{N} \to \widetilde{M},$$

$$\widetilde{N} = \left\{ (\lambda, t) \in N \mid t < t^1_{\text{Max}}(\lambda) \right\},$$

$$\widetilde{M} = M \setminus \text{cl}(\text{Max}).$$

This mapping may well be bijective, and if this is the case, then any point $q_1 \in \widetilde{M}$ is connected with q_0 by a unique candidate optimal geodesic; by virtue of existence, this geodesic is optimal.

The bijective property of the restricted exponential mapping can often be proved via the following classic theorem.

Theorem 3.11 (Hadamard) *Let* $F : X \to Y$ *be a smooth mapping between smooth manifolds for which the following conditions hold:*

(1) $\dim X = \dim Y$
(2) *X, Y are connected, and Y is simply connected*
(3) *F is nondegenerate*
(4) *F is proper (preimage of a compact set is compact).*

Then F is a diffeomorphism, thus a bijection.

Proof See [101, 103]. □

Usually it is difficult to describe all Maxwell points (and respectively to describe the first of them), but one can often do this for a group of symmetries G of the exponential mapping. Suppose that we have a mapping ε acting both in the preimage and image of the exponential mapping:

$$\varepsilon : N \to N, \qquad \varepsilon : M \to M.$$

This mapping is called a *symmetry of the exponential mapping* if it commutes with this mapping:

$$\varepsilon \circ \mathrm{Exp} = \mathrm{Exp} \circ \varepsilon$$

and if it preserves time:

$$\varepsilon(\lambda, t) = (*, t), \qquad (\lambda, t) \in N.$$

Suppose that there is a group G of symmetries of the exponential mapping. If

$$\varepsilon(\lambda, t) \neq (\lambda, t) \text{ and } \mathrm{Exp} \circ \varepsilon(\lambda, t) = \mathrm{Exp}(\lambda, t) = q_1,$$

$$\varepsilon \in G, \qquad (\lambda, t) \in N,$$

then q_1 is a Maxwell point. In such a way, one can describe the Maxwell points corresponding to the group of symmetries G, and consequently describe *the first Maxwell time corresponding to the group G*:

$$t_{\mathrm{Max}}^G : C \to (0, +\infty].$$

Then one can apply the above procedure with the restricted exponential mapping, replacing $t_{\mathrm{Max}}^1(\lambda)$ by $t_{\mathrm{Max}}^G(\lambda)$. If the group G is big enough, one can often prove that the restricted exponential mapping is bijective, and thus to construct optimal synthesis.

This method is an extension of the classic Hadamard's approach to the study of geodesics on Riemannian manifolds of negative curvature [101].

This symmetry method was successfully applied to a series of optimal control problems with a big symmetry group:

- Dido's problem (the sub-Riemannian problem on the Heisenberg group), see Sect. 4.1 and [37, 43, 79]
- the sub-Riemannian problem in the flat Martinet case, see Example 6 in Sect. 4.5 and [61]
- axisymmetric sub-Riemannian problems on the Lie groups SO(3), SU(2), SL(2), see [67, 68, 75]
- a general left-invariant sub-Riemannian problem on the Lie group SO(3), see [73]
- the sub-Riemannian problem with the growth vector (3, 6), see [83]
- the two-step sub-Riemannian problems of coranks 1 and 2, see [60, 66]
- the sub-Riemannian problem on the group of Euclidean motions of the plane, see Sect. 4.2 and [81, 91, 92]

- the sub-Riemannian problem on the group of hyperbolic motions of the plane, see [77]
- Euler's elastic problem, see Sect. 4.3 and [94]
- the problem on optimal rolling of a sphere on a plane without slipping, with twisting, see [72]
- the plate-ball problem, see [80, 90]
- sub-Riemannian problem on the Engel group, see Sect. 4.4 and [64]
- sub-Riemannian problem on the Cartan group, see Exercise 2 in Sect. 4.5 and [63, 96]
- axisymmetric Riemannian problems on the Lie groups SO(3), SU(2), SL(2), PSL(2), see [85, 86].

Chapter 4
Solution to Optimal Control Problems

The boy is not to separate himself with his whip and tether,
Lest the animal should wander away into a world of defilements;
When the ox is properly tended to, he will grow pure and docile;
Without a chain, nothing binding, he will by himself follow the oxherd.

Pu-ming, "The Ten Oxherding Pictures" (cited by Suzuki [110])

Herding the Ox

We present solutions to Dido's problem, Euler's elastic problem, and the sub-Riemannian problems on the group SE(2) and on the Engel group.

Yu. Sachkov, *Introduction to Geometric Control*, Springer Optimization
and Its Applications 192, https://doi.org/10.1007/978-3-031-02070-4_4

4.1   The Sub-Riemannian Problem on the Heisenberg Group

As we saw in Sect. 1.1.1.8, Dido's problem is stated as the following optimal control problem:

$$\dot{q} = u_1 f_1(q) + u_2 f_2(q), \qquad q \in M = \mathbb{R}^3_{x,y,z}, \quad u = (u_1, u_2) \in \mathbb{R}^2,$$

$$q(0) = q_0 = (0, 0, 0), \qquad q(t_1) = q_1,$$

$$J = \frac{1}{2} \int_0^{t_1} (u_1^2 + u_2^2)\, dt \to \min,$$

$$f_1 = \frac{\partial}{\partial x} - \frac{y}{2} \frac{\partial}{\partial z}, \quad f_2 = \frac{\partial}{\partial y} + \frac{x}{2} \frac{\partial}{\partial z}.$$

4.1.1   Existence of Solutions

We have $[f_1, f_2] = f_3 = \frac{\partial}{\partial z}$. The system is symmetric and full-rank, thus it is completely controllable.

The right-hand side satisfies the bound

$$|u_1 f_1(q) + u_2 f_2(q)| \le C(1 + |q|), \qquad q \in M, \quad u_1^2 + u_2^2 \le 1.$$

Thus the Filippov theorem gives existence of optimal controls.

4.1.2   Geodesics

Introduce linear on fibers of T^*M Hamiltonians:

$$h_i(\lambda) = \langle \lambda, f_i \rangle, \qquad i = 1, 2, 3, \quad \lambda \in T^*M.$$

4.1.2.1   Abnormal Case

Abnormal extremals satisfy the Hamiltonian system $\dot{\lambda} = u_1 \mathbf{h}_1(\lambda) + u_2 \mathbf{h}_2(\lambda)$, in coordinates:

$$\dot{h}_1 = -u_2 h_3,$$

$$\dot{h}_2 = u_1 h_3,$$

$$\dot{h}_3 = 0,$$

$$\dot{q} = u_1 f_1 + u_2 f_2,$$

plus the identities

$$h_1(\lambda_t) = h_2(\lambda_t) \equiv 0.$$

Thus $h_3(\lambda_t) \neq 0$, and the first two equations of the Hamiltonian system yield $u_1(t) = u_2(t) \equiv 0$. So abnormal trajectories are constant.

4.1.2.2   Normal Case

Normal extremals satisfy the Hamiltonian system $\dot{\lambda} = \mathbf{H}(\lambda)$ with the Hamiltonian $H = \frac{1}{2}(h_1^2 + h_2^2)$, in coordinates:

$$\dot{h}_1 = -h_2 h_3, \tag{4.1}$$

$$\dot{h}_2 = h_1 h_3, \tag{4.2}$$

$$\dot{h}_3 = 0, \tag{4.3}$$

$$\dot{q} = h_1 f_1 + h_2 f_2. \tag{4.4}$$

The subsystem of the Hamiltonian system for the adjoint variables h_1, h_2, h_3 (the *vertical subsystem*) (4.1)–(4.3) has integrals H and h_3. Moreover, in the plane $\{h_3 = 0\}$ the vertical subsystem stays fixed. Thus at the level surface $\{H = 1/2\}$ it has the flow shown in Fig. 4.1: rotations in the circles $\{H = 1/2, \ h_3 = \text{const} \neq 0\}$ and fixed points in the circle $\{H = 1/2, \ h_3 = 0\}$.

On the level surface $\{H = \frac{1}{2}\}$, we introduce the polar coordinate θ:

$$h_1 = \cos\theta, \quad h_2 = \sin\theta.$$

Fig. 4.1 Flow of the vertical subsystem (4.1)–(4.3)

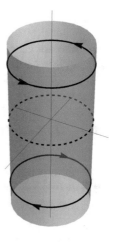

Arclength parametrized minimizers satisfy the normal Hamiltonian system

$$\dot\theta = h_3,$$
$$\dot h_3 = 0,$$
$$\dot x = \cos\theta,$$
$$\dot y = \sin\theta,$$
$$\dot z = -\frac{y}{2}\cos\theta + \frac{x}{2}\sin\theta,$$
$$(x, y, z)(0) = (0, 0, 0).$$

1. If $h_3 = 0$, then

$$\theta \equiv \theta_0,$$
$$x = t\cos\theta_0,$$
$$y = t\sin\theta_0,$$
$$z = 0.$$

In this case geodesics are lines in the plane $\{z = 0\}$, see Fig. 4.2.

2. If $h_3 \neq 0$, then

$$\theta = \theta_0 + h_3 t,$$
$$x = (\sin(\theta_0 + h_3 t) - \sin\theta_0)/h_3,$$
$$y = (\cos\theta_0 - \cos(\theta_0 + h_3 t))/h_3,$$
$$z = (h_3 t - \sin h_3 t)/h_3^2.$$

In this case geodesics are helices of nonconstant slope, they project to the plane (x, y) into circles, see Fig. 4.2.

4.1.3   Optimality of Geodesics

We present two ways to study optimality: an elementary one, specific for Dido's problem, and a general one, based on the symmetry method (see Sect. 3.4).

Fig. 4.2 Geodesics in Dido's problem

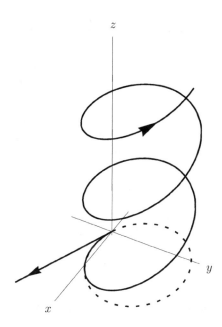

4.1.3.1   Elementary Argument

Straight lines (case $h_3 = 0$) minimize the Euclidean distance in $\mathbb{R}^2_{x,y}$, thus they are optimal on any segment $t \in [0, t_1]$, $t_1 > 0$.

Helices (case $h_3 \neq 0$) are not optimal after the first intersection with the z-axis at $t = \frac{2\pi}{|h_3|}$ since these intersections are Maxwell points. If $t_1 = \frac{2\pi}{|h_3|}$, then there is a continuum of helices $q(t)$, $t \in [0, t_1]$, coming to the same point $q(t_1)$ at the z-axis; they are obtained one from another by rotations around this axis, thus they all are optimal. A part of an optimal arc is optimal, thus the helices are optimal also for $t \in [0, t_1]$, $t_1 \in (0, \frac{2\pi}{|h_3|})$.

Summing up, the cut time along a geodesic $\mathrm{Exp}(\lambda, t)$ is

$$t_{\mathrm{cut}}(\lambda) = \begin{cases} \frac{2\pi}{|h_3|} & \text{for } h_3 \neq 0, \\ +\infty & \text{for } h_3 = 0. \end{cases} \qquad (4.5)$$

4.1.3.2   General Argument

As we mentioned above, in the case $h_3 = 0$ all geodesics are optimal.

In the case $h_3 \neq 0$ we study first the local optimality by evaluation of conjugate points:

$$J(t) = \frac{\partial \, \mathrm{Exp}}{\partial (\lambda_0, t)} = \frac{\partial (x, y, z)}{\partial (\theta_0, h_3, t)} = 0.$$

In the coordinates $p = \frac{h_3 t}{2}$, $\tau = \theta_0 + \frac{h_3 t}{2}$, we have:

$$x = \frac{2}{h_3} \cos \tau \sin p,$$

$$y = \frac{2}{h_3} \sin \tau \sin p,$$

$$z = \frac{2p - \sin 2p}{h_3^2}.$$

Thus

$$J = f \cdot \frac{\partial(x, y, z)}{\partial(\tau, p, h_3)} = f \cdot \frac{8 \sin p}{h_3^5} \cdot \varphi(p), \qquad f \neq 0,$$

$$\varphi(p) = (2p - \sin 2p) \cos p - (1 - \cos 2p) \sin p.$$

The function $\varphi(p)$ does not vanish for $p \in (0, \pi)$, thus the first root of the Jacobian J is $p_{\text{conj}}^1 = \pi$. Summing up, the first conjugate time in the case $h_3 \neq 0$ is

$$t_{\text{conj}}^1 = \frac{2\pi}{|h_3|}.$$

Now let us study the global optimality of geodesics. The problem has an obvious symmetry group G—rotations around the z-axis. The corresponding Maxwell times are $t = \frac{2\pi n}{h_3}$, and Maxwell points in the image of the exponential mapping are $x = y = 0$, $z = \frac{2\pi n}{h_3^2}$. The first Maxwell time corresponding to the group G is

$$t_{\text{Max}}^G = \frac{2\pi}{|h_3|} = t_{\text{conj}}^1.$$

Consider the restricted exponential mapping

$$\text{Exp} : \widetilde{N} \to \widetilde{M},$$

$$\widetilde{N} = \left\{ (\lambda, t) \in N \mid \theta \in S^1, \quad h_3 > 0, \quad t \in \left(0, \frac{2\pi}{h_3}\right) \right\},$$

$$\widetilde{M} = \{ q \in M \mid z > 0, \quad x^2 + y^2 > 0 \}.$$

The mapping $\text{Exp}\,|_{\widetilde{N}}$ is nondegenerate and proper:

$$(\theta, h_3, t) \to \partial \widetilde{N} \quad \Rightarrow \quad q \to \partial \widetilde{M}.$$

The manifolds $\widetilde{N}$, $\widetilde{M}$ are connected, but $\widetilde{M}$ is not simply connected. Thus the Hadamard theorem (see Theorem 3.11) cannot be applied immediately. In order to pass to a simply connected manifold, we factorize the exponential mapping by the group of rotations. We get

$$\widehat{N} = \widetilde{N}/S^1 = \left\{ (h_3, t) \in \mathbb{R}^2 \mid h_3 > 0, \quad t \in \left(0, \frac{2\pi}{h_3} \right) \right\},$$

$$\widehat{M} = \widetilde{M}/S^1 = \left\{ (r, z) \in \mathbb{R}^2 \mid z > 0, \quad r = \sqrt{x^2 + y^2} > 0 \right\},$$

$$\widehat{\mathrm{Exp}} : \widehat{N} \to \widehat{M}, \qquad \widehat{\mathrm{Exp}}(h_3, t) = (z, r),$$

$$z = \frac{2p - \sin 2p}{h_3^2}, \quad r = \frac{2}{h_3} \sin p, \quad p = \frac{h_3 t}{2}.$$

By the Hadamard theorem (see Theorem 3.11), the mapping $\widehat{\mathrm{Exp}} : \widehat{N} \to \widehat{M}$ is a diffeomorphism, thus $\mathrm{Exp} : \widetilde{N} \to \widetilde{M}$ is a diffeomorphism as well.

So for any $q_1 \in \widetilde{M}$ there exists a unique $(\lambda_0, t_1) \in \widetilde{N}$ such that $q_1 = \mathrm{Exp}(\lambda_0, t_1)$. Thus the geodesic $q(t) = \mathrm{Exp}(\lambda_0, t)$, $t \in [0, t_1]$, is optimal. Summing up, if $z_1 \neq 0$, $x_1^2 + y_1^2 \neq 0$, then there exists a unique minimizer connecting q_0 with $q_1 = (x_1, y_1, z_1)$ determined by parameters $\theta_0 \in S^1$, $h_3 \neq 0$, $t_1 \in (0, \frac{2\pi}{|h_3|})$.

If $z_1 = 0$, $x_1^2 + y_1^2 \neq 0$, then there exists a unique minimizer determined by parameters $\theta_0 \in S^1$, $h_3 = 0$, $t_1 > 0$.

Finally, if $z_1 \neq 0$, $x_1^2 + y_1^2 = 0$, then there exists a one-parameter family of minimizers determined by parameters $\theta_0 \in S^1$, $h_3 \neq 0$, $t_1 = \frac{2\pi}{|h_3|}$.

Once more we recover expression (4.5) for the cut time in Dido's problem.

Some optimal trajectories starting at the point q_0 with the same tangent vector are shown in Fig. 4.3.

Let us return to the initial statement of Dido's problem in Sect. 1.1.1.8. Let the curve γ_0 be a line segment. If $S = 0$, then the optimal curve γ is the same line segment passed in the opposite direction. And if $S \neq 0$, then γ is an arc of a circle (a complete circle passed once if $a_1 = a_0$, or its part if $a_1 \neq a_0$).

4.1.4   Cut Locus and Caustic

In Dido's problem the *cut locus*

$$\mathrm{Cut} = \{ \mathrm{Exp}(\lambda, t_{\mathrm{cut}}(\lambda)) \mid \lambda \in C \}$$

and the first *caustic*

$$\mathrm{Conj}^1 = \left\{ \mathrm{Exp}(\lambda, t_{\mathrm{conj}}^1(\lambda)) \mid \lambda \in C \right\}$$

Fig. 4.3 Optimal trajectories
in Dido's problem

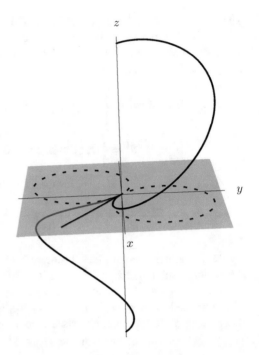

coincides one with another:

$$\text{Cut} = \text{Conj}^1 = \{(0, 0, z) \in \mathbb{R}^3 \mid z \neq 0\}.$$

4.1.5 Sub-Riemannian Distance and Spheres

Let us describe the SR distance $d_0(q) = d(q_0, q)$, $q = (x, y, z) \in \mathbb{R}^3$:

- if $z = 0$, then $d_0(q) = \sqrt{x^2 + y^2}$,
- if $z \neq 0$, $x^2 + y^2 = 0$, then $d_0(q) = \sqrt{2\pi |z|}$,
- if $z \neq 0$, $x^2 + y^2 \neq 0$, then the distance is determined by the conditions

$$d_0(q) = \frac{p}{\sin p}\sqrt{x^2 + y^2},$$

$$\frac{2p - \sin 2p}{4 \sin^2 p} = \frac{z}{x^2 + y^2}.$$

The unit *sub-Riemannian sphere* $S = \{q \in \mathbb{R}^3 \mid d_0(q) = 1\}$ is a surface of revolution around the axis z in the form of an apple, see Figs. 4.4 and 4.5. It has two singular conical points $z = \pm\frac{1}{4\pi}$, $x^2 + y^2 = 0$. The remaining spheres $S_R = \{q \in$

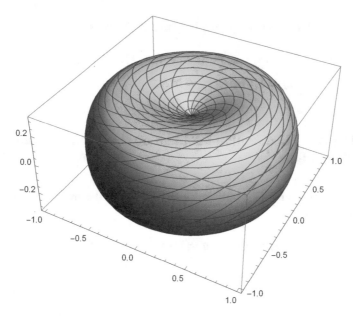

Fig. 4.4 Sub-Riemannian sphere on the Heisenberg group

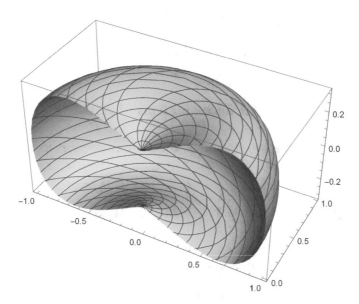

Fig. 4.5 Sub-Riemannian hemisphere on the Heisenberg group

$\mathbb{R}^3 \mid d_0(q) = R\}$ are obtained from S by virtue of dilations:

$$\delta_s : (x, y, z) \mapsto (e^s x, e^s y, e^{2s} z), \qquad s \in \mathbb{R},$$
$$S_R = \delta_s(S), \qquad s = \ln R.$$

4.2   The Sub-Riemannian Problem on the Group of Motions of the Plane

Recall the statement of the sub-Riemannian problem on the group of motions of the plane, see Sect. 1.1.1:

$$\dot{x} = u_1 \cos \theta, \qquad q = (x, y, \theta) \in M = \mathbb{R}^2 \times S^1, \tag{4.6}$$

$$\dot{y} = u_1 \sin \theta, \qquad u = (u_1, u_2) \in \mathbb{R}^2, \tag{4.7}$$

$$\dot{\theta} = u_2, \tag{4.8}$$

$$q(0) = q_0, \qquad q(t_1) = q_1, \tag{4.9}$$

$$l = \int_0^{t_1} \sqrt{u_1^2 + u_2^2}\, dt \to \min. \tag{4.10}$$

In this section we solve problem (4.6)–(4.10).

4.2.1   The Group of Euclidean Motions of the Plane

Proper motions of the Euclidean plane (rotations and parallel translations) are parametrised by matrices of the form

$$Q = \begin{pmatrix} \cos \theta & -\sin \theta & x \\ \sin \theta & \cos \theta & y \\ 0 & 0 & 1 \end{pmatrix}, \qquad \theta \in S^1, \quad (x, y) \in \mathbb{R}^2. \tag{4.11}$$

The action of a motion Q on a vector $(a, b) \in \mathbb{R}^2$ is given as follows:

$$Q(a, b) = \begin{pmatrix} \cos \theta & -\sin \theta & x \\ \sin \theta & \cos \theta & y \\ 0 & 0 & 1 \end{pmatrix} \begin{pmatrix} a \\ b \\ 1 \end{pmatrix} = \begin{pmatrix} a \cos \theta - b \sin \theta + x \\ a \sin \theta + b \cos \theta + y \\ 1 \end{pmatrix},$$

and is the rotation by the angle θ around the origin and the parallel translation by the vector (x, y).

Thus the *group of proper motions of the two-dimensional Euclidean plane* has a representation

$$SE(2) = \left\{ \begin{pmatrix} \cos\theta & -\sin\theta & x \\ \sin\theta & \cos\theta & y \\ 0 & 0 & 1 \end{pmatrix} \middle| \; \theta \in S^1, \; (x, y) \in \mathbb{R}^2 \right\},$$

and is a 3-dimensional linear Lie group.

4.2.2   The Left-Invariant Sub-Riemannian Problem on SE(2)

We show that problem (4.6)–(4.10) is a left-invariant problem on the Lie group SE(2).

Differentiating matrix (4.11) w.r.t. system (4.6)–(4.8), we get

$$\dot{Q} = \begin{pmatrix} -\dot\theta\sin\theta & -\dot\theta\cos\theta & \dot{x} \\ \dot\theta\cos\theta & -\dot\theta\sin\theta & \dot{y} \\ 0 & 0 & 0 \end{pmatrix} = Q \begin{pmatrix} 0 & -u_2 & u_1 \\ u_2 & 0 & 0 \\ 0 & 0 & 0 \end{pmatrix}.$$

Thus problem (4.6)–(4.10) is rewritten in the form

$$\dot{Q} = Q \begin{pmatrix} 0 & -u_2 & u_1 \\ u_2 & 0 & 0 \\ 0 & 0 & 0 \end{pmatrix}, \qquad Q \in SE(2), \quad (u_1, u_2) \in \mathbb{R}^2, \tag{4.12}$$

$$Q(0) = Q_0, \qquad Q(t_1) = Q_1, \tag{4.13}$$

$$l = \int_0^{t_1} \sqrt{u_1^2 + u_2^2}\, dt \to \min. \tag{4.14}$$

Consequently, problem (4.12)–(4.14), and an isomorphic problem (4.6)–(4.10) is a left-invariant sub-Riemannian problem on the Lie group SE(2). So we can assume that the starting point is the identity of the group: $Q_0 = \mathrm{Id}$, i.e., $q_0 = (0, 0, 0)$.

4.2.3   Existence of Solutions

Let us rewrite system (4.6)–(4.8) in the vector form:

$$\dot{q} = u_1 f_1(q) + u_2 f_2(q), \qquad q \in M = \mathbb{R}^2 \times S^1, \tag{4.15}$$

$$f_1 = \cos\theta \frac{\partial}{\partial x} + \sin\theta \frac{\partial}{\partial y}, \qquad f_2 = \frac{\partial}{\partial\theta}. \tag{4.16}$$

As we noticed in Sect. 1.1.1, this system is completely controllable by the Rashevskii–Chow theorem.

The right-hand side of system (4.15), after transformation of the SR problem to an equivalent time-optimal problem with controls $u_1^2 + u_2^2 \leq 1$, satisfies the inequality $|u_1 f_1 + u_2 f_2| \leq C$, which implies an a priori bound for the attainable set. By the Filippov theorem, optimal control exists.

4.2.4   The Pontryagin Maximum Principle

Introduce linear on fibers of T^*M Hamiltonians $h_i(\lambda) = \langle \lambda, f_i(q) \rangle$, $\lambda \in T^*M$, $q = \pi(\lambda)$, $f_3 = [f_1, f_2] = \sin\theta \frac{\partial}{\partial x} - \cos\theta \frac{\partial}{\partial y}$, and the Hamiltonian of PMP:

$$h_u^\nu(\lambda) = u_1 h_1(\lambda) + u_2 h_2(\lambda) + \frac{\nu}{2}(u_1^2 + u_2^2).$$

The corresponding Hamiltonian system has the form

$$\dot{h}_1 = \{u_1 h_1 + u_2 h_2, h_1\} = -u_2 h_3,$$
$$\dot{h}_2 = \{u_1 h_1 + u_2 h_2, h_2\} = u_1 h_3,$$
$$\dot{h}_3 = \{u_1 h_1 + u_2 h_2, h_3\} = u_2 h_1,$$
$$\dot{q} = u_1 f_1 + u_2 f_2.$$

4.2.4.1   Abnormal Case

Let $\nu = 0$. The maximality condition of PMP

$$u_1 h_1 + u_2 h_2 \rightarrow \max_{(u_1, u_2) \in \mathbb{R}^2}$$

gives

$$h_1 = h_2 = 0 \quad \Rightarrow \quad h_3 \neq 0,$$

then the Hamiltonian system yields

$$0 = \dot{h}_1 = -u_2 h_3 \quad \Rightarrow \quad u_2 \equiv 0,$$
$$0 = \dot{h}_2 = u_1 h_3 \quad \Rightarrow \quad u_1 \equiv 0,$$

i.e., the abnormal trajectory is trivial: $q(t) \equiv q_0$.

4.2.4.2   Normal Case

Let $\nu = -1$. The maximality condition of PMP

$$u_1 h_1 + u_2 h_2 - \frac{1}{2}(u_1^2 + u_2^2) \to \max_{(u_1, u_2) \in \mathbb{R}^2}$$

yields

$$u_1 = h_1, \qquad u_2 = h_2,$$

and the maximized normal Hamiltonian is $H = \frac{1}{2}(h_1^2 + h_2^2)$. The corresponding Hamiltonian system $\dot{\lambda} = \mathbf{H}(\lambda)$ has the form

$$\dot{h}_1 = -h_2 h_3, \qquad \dot{h}_2 = h_1 h_3, \qquad \dot{h}_3 = h_2 h_1, \qquad (4.17)$$

$$\dot{x} = h_1 \cos\theta, \qquad \dot{y} = h_1 \sin\theta, \qquad \dot{\theta} = h_2. \qquad (4.18)$$

This system has integrals: the Hamiltonian H and $F = h_1^2 + h_3^2$. The flow of the vertical subsystem (4.17) in the level surface $\{H = \frac{1}{2}\}$ is obtained by intersecting this surface with the cylinders $F = \text{const}$, see Fig. 4.6.

Different qualitative types of intersections of the level surface $\{H = \frac{1}{2}\}$ with the cylinders $F = \text{const}$ are shown in Figs. 4.7, 4.8, 4.9 and 4.10.

On the level surface $\{H = \frac{1}{2}\}$, introduce the polar coordinate: $h_1 = \cos\alpha$, $h_2 = \sin\alpha$, then the vertical subsystem (4.17) of the normal Hamiltonian system takes the form

$$\dot{\alpha} = h_3,$$

$$\dot{h}_3 = \frac{1}{2}\sin 2\alpha.$$

Fig. 4.6 Intersection of level surfaces of integrals H and F

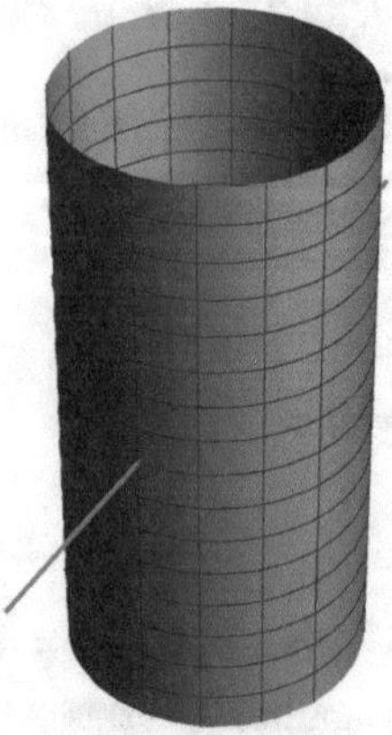

Fig. 4.7 Intersection of
$\{H = 1/2\}$ and $\{F = 0\}$

Fig. 4.8 Intersection of
$\{H = 1/2\}$ and
$\{F = \text{const} \in (0, 1)\}$

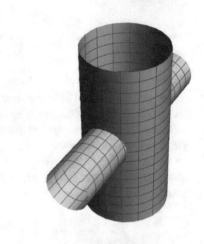

Fig. 4.9 Intersection of
$\{H = 1/2\}$ and $\{F = 1\}$

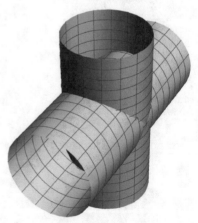

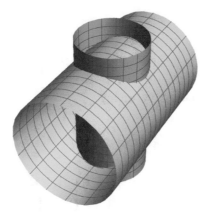

Fig. 4.10 Intersection of $\{H = 1/2\}$ and $\{F = \text{const} > 1\}$

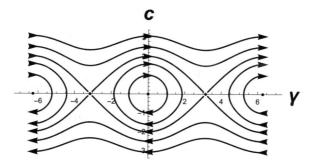

Fig. 4.11 Phase portrait of pendulum (4.19) and (4.20)

Further, define the variables

$$\gamma = 2\alpha + \pi, \qquad c = 2h_3,$$

then the vertical subsystem (4.17) turns into a (double covering of a) pendulum

$$\dot{\gamma} = c, \qquad\qquad\qquad \gamma \in \mathbb{R}/(4\pi\mathbb{Z}), \qquad\qquad (4.19)$$

$$\dot{c} = -\sin\gamma, \qquad\qquad\qquad c \in \mathbb{R}. \qquad\qquad\qquad (4.20)$$

See its phase portrait in Fig. 4.11.

Pendulum (4.19) and (4.20) has a full energy integral

$$E = 2(F - H) = 2h_3^2 + h_1^2 - h_2^2 = \frac{c^2}{2} - \cos\gamma \equiv \text{const} \in [-1, \infty).$$

The phase cylinder of the pendulum

$$C = T_{q_0}^* M \bigcap \left\{ H = \frac{1}{2} \right\}$$

stratifies according to values of the energy:

$$C = \bigsqcup_{i=1}^{5} C_i, \tag{4.21}$$

$$C_1 = \{\lambda \in C \mid E \in (-1, 1)\},$$
$$C_2 = \{\lambda \in C \mid E \in (1, \infty)\},$$
$$C_3 = \{\lambda \in C \mid E = 1,\ c \neq 0\},$$
$$C_4 = \{\lambda \in C \mid E = -1\},$$
$$C_5 = \{\lambda \in C \mid E = 1,\ c = 0\}.$$

See stratification (4.21) in Fig. 4.12.

Periodic trajectories of pendulum (4.19) and (4.20) have the following periods:

$$\lambda \in C_1 \quad \Rightarrow \quad T = 4K(k), \quad k = \sqrt{\frac{E+1}{2}},$$

$$\lambda \in C_2 \quad \Rightarrow \quad T = 4kK(k), \quad k = \sqrt{\frac{2}{E+1}},$$

where $K(k)$ is the complete elliptic integral of the first kind; see Sect. A.1.1.

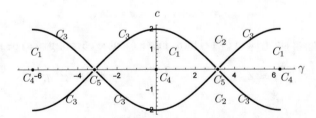

Fig. 4.12 Stratification (4.21) of the phase cylinder of pendulum (4.19) and (4.20)

4.2.5   Geodesics and the Exponential Mapping

Compute the curvature κ of projections of geodesics to the plane (x, y). Taking into account the normal Hamiltonian system (4.17) and (4.18), we get

$$\kappa = \frac{\dot{x}\ddot{y} - \ddot{x}\dot{y}}{(\dot{x}^2 + \dot{y}^2)^{\frac{3}{2}}} = \frac{h_2}{|h_1|} = -\frac{\cos(\gamma/2)}{|\sin(\gamma/2)|}.$$

A curve $(x(t), y(t))$ has an inflection point if and only if $\kappa = 0$, i.e., $\gamma = \pi + 2\pi n$. A curve $(x(t), y(t))$ has a cusp point if and only if $\kappa = \infty$, i.e., $\gamma = 2\pi n$. Plots of curves $(x(t), y(t))$ for $\lambda \in \overset{3}{\underset{i=1}{\cup}} C_i$ are given in Figs. 4.13, 4.14 and 4.15.

In the case $\lambda \in C_4$ the curve $(x(t), y(t))$ is constant (the car rotates at the place), and in the case $\lambda \in C_5$ it is a straight line (the car moves forward).

Fig. 4.13 Non-inflectional trajectory in SE(2): $\lambda \in C_1$

Fig. 4.14 Inflectional trajectory in SE(2): $\lambda \in C_2$

Fig. 4.15 Critical trajectory in SE(2): $\lambda \in C_3$

The family of all geodesics is parametrized by the exponential mapping

$$\text{Exp} \ : \ N = C \times \mathbb{R}_+ \rightarrow M, \qquad (\lambda_0, t) \mapsto \pi \circ e^{t\mathbf{H}}(\lambda_0) = q(t).$$

4.2.6   Symmetries of the Exponential Mapping

4.2.6.1   Reflections in the Preimage of the Exponential Mapping

The phase portrait of pendulum (4.19) and (4.20) admits the following reflections:

$$\varepsilon^1 \ : \ (\gamma, c) \mapsto (\gamma_1, c_1) = (\gamma, -c), \tag{4.22}$$

$$\varepsilon^2 \ : \ (\gamma, c) \mapsto (\gamma_2, c_2) = (-\gamma, c), \tag{4.23}$$

$$\varepsilon^3 \ : \ (\gamma, c) \mapsto (\gamma_3, c_3) = (-\gamma, -c), \tag{4.24}$$

$$\varepsilon^4 \ : \ (\gamma, c) \mapsto (\gamma_4, c_4) = (\gamma + 2\pi, c), \tag{4.25}$$

$$\varepsilon^5 \ : \ (\gamma, c) \mapsto (\gamma_5, c_5) = (\gamma + 2\pi, -c), \tag{4.26}$$

$$\varepsilon^6 \ : \ (\gamma, c) \mapsto (\gamma_6, c_6) = (-\gamma + 2\pi, c), \tag{4.27}$$

$$\varepsilon^7 \ : \ (\gamma, c) \mapsto (\gamma_7, c_7) = (-\gamma + 2\pi, -c), \tag{4.28}$$

see Fig. 4.16.

These reflections generate the group of symmetries of a parallelepiped $G = \{\text{Id}, \varepsilon^1, \dots, \varepsilon^7\}$. The reflections ε^3, ε^4, ε^7 preserve the direction of time on trajectories of the pendulum, while the rest reflections ε^1, ε^2, ε^5, ε^6 reverse the direction of time.

Proposition 4.1 *The following mappings transform trajectories of pendulum (4.19) and (4.20) to trajectories:*

$$\varepsilon^i \ : \ \{(\gamma_s, c_s) \mid s \in [0, t]\} \mapsto \{(\gamma_s^i, c_s^i) \mid s \in [0, t]\}, \qquad i = 1, \dots, 7, \tag{4.29}$$

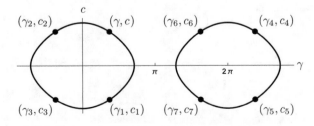

Fig. 4.16 Reflections (4.22)–(4.28) in the phase cylinder of pendulum

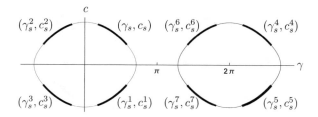

Fig. 4.17 Reflections (4.30)–(4.36) of trajectories of pendulum

where

$$(\gamma_s^1, c_s^1) = (\gamma_{t-s}, -c_{t-s}),$$ (4.30)

$$(\gamma_s^2, c_s^2) = (-\gamma_{t-s}, c_{t-s}),$$ (4.31)

$$(\gamma_s^3, c_s^3) = (-\gamma_s, -c_s),$$ (4.32)

$$(\gamma_s^4, c_s^4) = (\gamma_s + 2\pi, c_s),$$ (4.33)

$$(\gamma_s^5, c_s^5) = (\gamma_{t-s} + 2\pi, -c_{t-s}),$$ (4.34)

$$(\gamma_s^6, c_s^6) = (-\gamma_{t-s} + 2\pi, c_{t-s}),$$ (4.35)

$$(\gamma_s^7, c_s^7) = (-\gamma_s + 2\pi, -c_s),$$ (4.36)

see Fig. 4.17.

Proof The statement is verified by substitution to system (4.19), (4.20) and differentiation.   □

The *reflections in the preimage of the exponential mapping N*

$$\varepsilon^i : (\lambda, t) \mapsto (\lambda^i, t), \qquad i = 1, \dots, 7,$$

are defined as follows. Let $(\lambda, t) = (\gamma, c, t) \in N$ and let (γ_s, c_s), $s \in [0, t]$, be a solution to the equation of pendulum (4.19) and (4.20) with the initial condition $(\gamma_0, c_0) = (\gamma, c)$. Then

$$\varepsilon^i(\lambda, t) = (\lambda^i, t) = (\gamma^i, c^i, t) \in N,$$

where

$$(\gamma^1, c^1) = (\gamma_t, -c_t),$$ (4.37)

$$(\gamma^2, c^2) = (-\gamma_t, c_t),$$ (4.38)

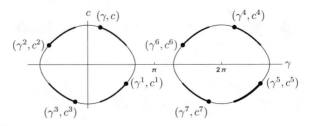

Fig. 4.18 Reflections (4.37)–(4.43)

$$(\gamma^3, c^3) = (-\gamma, -c), \tag{4.39}$$

$$(\gamma^4, c^4) = (\gamma + 2\pi, c), \tag{4.40}$$

$$(\gamma^5, c^5) = (\gamma_t + 2\pi, -c_t), \tag{4.41}$$

$$(\gamma^6, c^6) = (2\pi - \gamma_t, c_t), \tag{4.42}$$

$$(\gamma^7, c^7) = (2\pi - \gamma, -c), \tag{4.43}$$

see Fig. 4.18.

These reflections generate the same group of symmetries of the parallelepiped:

$$G = \{\mathrm{Id}, \varepsilon^1, \dots, \varepsilon^7\} \cong \mathbb{Z}_2 \times \mathbb{Z}_2 \times \mathbb{Z}_2,$$

where $\mathbb{Z}_2 = \mathbb{Z}/(2\mathbb{Z})$.

4.2.6.2 Reflections in the Image of the Exponential Mapping

We define action of the group G on the normal extremals $\lambda_s = e^{s\mathbf{H}}(\lambda_0) \in T^*M$, $s \in [0, t]$, i.e., solutions to the normal Hamiltonian system

$$\dot{\gamma}_s = c_s, \qquad \dot{c}_s = -\sin\gamma_s, \tag{4.44}$$

$$\dot{q}_s = \sin\frac{\gamma_s}{2} f_1(q_s) - \cos\frac{\gamma_s}{2} f_2(q_s), \tag{4.45}$$

as follows:

$$\varepsilon^i : \{\lambda_s \mid s \in [0, t]\} \mapsto \{\lambda_s^i \mid s \in [0, t]\}, \qquad i = 1, \dots, 7, \tag{4.46}$$

$$\lambda_s = (\gamma_s, c_s, q_s), \qquad \lambda_s^i = (\gamma_s^i, c_s^i, q_s^i). \tag{4.47}$$

Here λ_s^i is a solution to the Hamiltonian system (4.44) and (4.45), and the action of reflections on the vertical coordinates (γ_s, c_s) was defined in Proposition 4.1.

The action of reflections on the horizontal coordinates (x_s, y_s, θ_s) is described as follows.

Proposition 4.2 *Let* $q_s = (x_s, y_s, \theta_s)$, $s \in [0, t]$, *be a normal extremal trajectory, and let* $q_s^i = (x_s^i, y_s^i, \theta_s^i)$, $s \in [0, t]$, *be its image under the action of the reflection* ε^i, $i = 1, \ldots, 7$, *as defined by (4.46) and (4.47). Then the following equalities hold:*

(1)
$$\theta_1^s = \theta_t - \theta_{t-s},$$
$$x_s^1 = \cos \theta_t (x_t - x_{t-s}) + \sin \theta_t (y_t - y_{t-s}),$$
$$y_s^1 = \sin \theta_t (x_t - x_{t-s}) - \cos \theta_t (y_t - y_{t-s}),$$

(2)
$$\theta_2^s = \theta_t - \theta_{t-s},$$
$$x_s^2 = - \cos \theta_t (x_t - x_{t-s}) - \sin \theta_t (y_t - y_{t-s}),$$
$$y_s^2 = - \sin \theta_t (x_t - x_{t-s}) + \cos \theta_t (y_t - y_{t-s}),$$

(3)
$$\theta_3^s = \theta_s,$$
$$x_s^3 = -x_s,$$
$$y_s^3 = -y_s,$$

(4)
$$\theta_4^s = -\theta_s,$$
$$x_s^4 = -x_s,$$
$$y_s^4 = y_s,$$

(5)
$$\theta_5^s = \theta_{t-s} - \theta_t,$$
$$x_s^5 = \cos \theta_t (x_{t-s} - x_t) + \sin \theta_t (y_{t-s} - y_t),$$
$$y_s^5 = - \sin \theta_t (x_{t-s} - x_t) + \cos \theta_t (y_{t-s} - y_t),$$

(6)
$$\theta_6^s = \theta_{t-s} - \theta_t,$$
$$x_s^6 = \cos \theta_t (x_t - x_{t-s}) + \sin \theta_t (y_t - y_{t-s}),$$
$$y_s^6 = - \sin \theta_t (x_t - x_{t-s}) + \cos \theta_t (y_t - y_{t-s}),$$

(7)
$$\theta_7^s = -\theta_s,$$
$$x_s^7 = x_s,$$
$$y_s^7 = -y_s.$$

Proof Integration of ODE (4.45), with the use of Proposition 4.1.   □

Modulo rotations in the plane (x, y), the action of the reflections ε^i on curves $(x(t), y(t))$ has the following visual meaning:

Fig. 4.19 Action of ε^1, ε^2

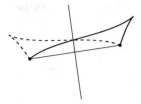

Fig. 4.20 Action of ε^4, ε^7

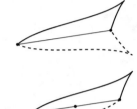

Fig. 4.21 Action of ε^5, ε^6

- ε^1, ε^2 are reflections in the middle perpendicular to the chord of a curve $(x(t), y(t))$; see Fig. 4.19,
- ε^4, ε^7 are reflections in this chord; see Fig. 4.20,
- ε^5, ε^6 are reflections in the centre of this chord; see Fig. 4.21.

Further, we define the action of *reflections in the image of the exponential mapping M* as the action on endpoints of extremal trajectories

$$\varepsilon^i \ : \ M \to M, \qquad \varepsilon^i \ : \ q_t \mapsto q_t^i, \tag{4.48}$$

see (4.46) and (4.47). By virtue of Proposition 4.2, the point q_t^i depends only on the endpoint q_t, not on the whole trajectory $\{q_s \mid s \in [0, t]\}$.

Proposition 4.3 *Let* $q = (x, y, \theta) \in M$, $q^i = \varepsilon^i(q) = (x^i, y^i, \theta^i) \in M$, $i = 1, \ldots, 7$. *Then:*

$$(x^1, y^1, \theta^1) = (x \cos\theta + y \sin\theta, x \sin\theta - y \cos\theta, \theta),$$

$$(x^2, y^2, \theta^2) = (-x \cos\theta - y \sin\theta, -x \sin\theta + y \cos\theta, \theta),$$

$$(x^3, y^3, \theta^3) = (-x, -y, \theta),$$

$$(x^4, y^4, \theta^4) = (-x, y, -\theta),$$

$$(x^5, y^5, \theta^5) = (-x \cos\theta - y \sin\theta, x \sin\theta - y \cos\theta, -\theta),$$

$$(x^6, y^6, \theta^6) = (x \cos\theta + y \sin\theta, -x \sin\theta + y \cos\theta, -\theta),$$

$$(x^7, y^7, \theta^7) = (x, -y, -\theta).$$

Proof It suffices to substitute $s = 0$ to the formulas of Proposition 4.2. $\qquad\square$

4.2.7 Maxwell Time Corresponding to Symmetries

Define *the first Maxwell time corresponding to the i-th symmetry*

$$t_{\text{Max}}^i(\lambda) = \inf\{t > 0 \mid \varepsilon^i \circ \text{Exp}(\lambda, t) = \text{Exp}(\lambda, t), \ \varepsilon^i(\lambda, t) \neq (\lambda, t)\},$$
$$i = 1, \dots, 7, \quad \lambda \in C,$$

and *the first Maxwell time corresponding to the group G*:

$$t_{\text{Max}}^G(\lambda) = \min\{t_{\text{Max}}^i(\lambda) \mid i = 1, \dots, 7\}, \qquad \lambda \in C.$$

In paper [81] the following expression for this time was obtained:

$$t_{\text{Max}}^G(\lambda) = \begin{cases} 2K(k) = T/2, & \lambda \in C_1, \\ 2kp_1(k) \in (T/2, T), & \lambda \in C_2, \\ \pi, & \lambda \in C_4, \\ +\infty, & \lambda \in C_3 \cup C_5, \end{cases}$$

where $p = p_1(k) \in (K(k), 2K(k))$ is the first positive root of the equation

$$f(p, k) = \text{cn}\, p(\text{E}(p) - p) - \text{dn}\, p \, \text{sn}\, p = 0,$$

and cn, sn, dn are Jacobi's elliptic functions, E is Jacobi's epsilon function; see Sect. A.1; T is the period of oscillations of pendulum (4.19) and (4.20).

Let us bound the *cut time*

$$t_{\text{cut}}(\lambda) = \sup\{t_1 > 0 \mid \text{Exp}(\lambda, t) \text{ is optimal for } t \in [0, t_1]\}.$$

On the basis of the study of Maxwell points corresponding to the group G, we have the following upper bound of the cut time.

Theorem 4.1 *For any $\lambda \in C$ we have $t_{\text{cut}}(\lambda) \leq t_{\text{Max}}^G(\lambda)$.*

4.2.8 Conjugate Points

Consider *the first conjugate time* along a normal geodesic $\text{Exp}(\lambda, t)$, $\lambda \in C$:

$$t_{\text{conj}}^1(\lambda) = \inf\{t > 0 \mid \text{Exp}_{*(\lambda, t)} \text{ is degenerate}\}.$$

Theorem 4.2 *If $\lambda \in C_1 \cup C_3 \cup C_4 \cup C_5$, then $t_{\text{conj}}^1(\lambda) = +\infty$.*

If $\lambda \in C_2$, then $t_{\text{conj}}^1(\lambda) \in [t_{\text{Max}}^G(\lambda), 4kK(k)]$.

Proof See [91]. □

Theorem 4.2 is proved via homotopy invariance of index of the second variation (number of conjugate points with account of multiplicity). An extremal corresponding to some motion of the pendulum is continuously deformed to an extremal corresponding to a motion of the pendulum near the stable equilibrium ($k \to 0$), where the elliptic functions degenerate:

$$\text{cn}(\varphi, k) \to \cos\varphi, \qquad\qquad \text{sn}(\varphi, k) \to \sin\varphi,$$

$$\text{dn}(\varphi, k) \to 1, \qquad\qquad \text{E}(\varphi, k) \to \varphi.$$

Corollary 4.1 *For any $\lambda \in C$ we have $t_{\text{conj}}^1(\lambda) \geq t_{\text{Max}}^G(\lambda)$.*

Thus the geodesic $\text{Exp}(\lambda, t)$ is locally optimal for $t \in [0, t_{\text{Max}}^G(\lambda)]$.

4.2.9 Structure of the Exponential Mapping

The exponential mapping

$$\text{Exp} : N \to M, \qquad N = C \times \mathbb{R}_+,$$

has multiple points (Maxwell points), thus it is not bijective. Let us restrict it to subsets without Maxwell points:

$$\text{Exp} : \widetilde{N} \to \widetilde{M},$$

$$\widetilde{N} = \{(\lambda, t) \in N \mid t < t_{\text{Max}}^G(\lambda)\},$$

$$\widetilde{M} = M \setminus \{q \in M \mid \exists i = 1, \dots, 7 \text{ s.t. } \varepsilon^i(q) = q\}.$$

The restricted exponential mapping $\text{Exp} : \widetilde{N} \to \widetilde{M}$ is injective.

One can prove that it is bijective via Hadamard's theorem (see Theorem 3.11). With the help of this theorem the following statement was proved [91].

Theorem 4.3 *The restricted exponential mapping $\text{Exp} : \widetilde{N} \to \widetilde{M}$ is a diffeomorphism, thus a bijection.*

Corollary 4.2 *For any point $q_1 \in \widetilde{M}$ denote the preimage $\text{Exp}^{-1}(q_1) = (\lambda_0, t_1) \in \widetilde{N}$. Then the geodesic $\text{Exp}(\lambda_0, t)$ is optimal for $t \in [0, t_1]$.*

Corollary 4.3 *For any $\lambda \in C$ we have $t_{\text{cut}}(\lambda) = t_{\text{Max}}^G(\lambda)$.*

4.2.10   Cut Locus and Caustic

The *cut locus* is defined as

$$\mathrm{Cut} = \{\mathrm{Exp}(\lambda, t_{\mathrm{cut}}(\lambda)) \mid \lambda \in C\}.$$

It was proved [92] that in the SR problem on SE(2) the cut locus has 3 connected components:

$$\mathrm{Cut} = \mathrm{Cut}_{\mathrm{loc}}^{+} \cup \mathrm{Cut}_{\mathrm{loc}}^{-} \cup \mathrm{Cut}_{\mathrm{glob}},$$

$$q_0 \in \mathrm{cl}(\mathrm{Cut}_{\mathrm{loc}}^{\pm}),$$

$$q_0 \text{ is isolated from } \mathrm{Cut}_{\mathrm{glob}},$$

see Fig. 4.22. The *global component of the cut locus* is

$$\mathrm{Cut}_{\mathrm{glob}} = \{q \in M \mid \theta = \pi\},$$

it is depicted by the disk in Fig. 4.22. The *local components of the cut locus* are contained in a Möbius strip:

$$\mathrm{Cut}_{\mathrm{loc}}^{\pm} \subset \{q \in M \mid R(q) = 0\},$$

$$R(q) = x \cos\frac{\theta}{2} + y \sin\frac{\theta}{2},$$

they are depicted by parts of the Möbius strip $\{R(q) = 0\}$ in Fig. 4.22.

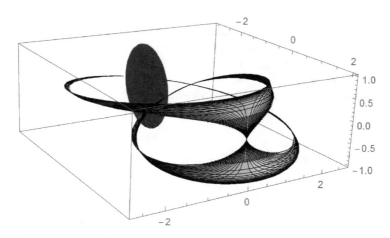

Fig. 4.22 Cut locus Cut $\subset$ SE(2)

Fig. 4.23 Sub-Riemannian
caustic on SE(2)

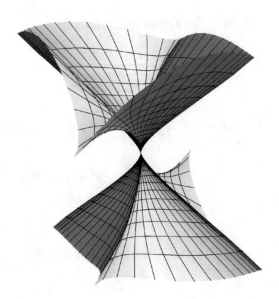

Figure 4.23 shows the first *sub-Riemannian caustic*

$$\text{Conj}^1 = \left\{ \text{Exp}(\lambda, t_{\text{conj}}^1(\lambda)) \mid \lambda \in C) \right\}.$$

The optimal synthesis in the problem was constructed in papers [81, 91, 92].

4.2.11   Explicit Optimal Solutions for Special Terminal Points

In this subsection we describe optimal solutions $q_t = (x_t, y_t, \theta_t) = \text{Exp}(\lambda, t)$, $t \in C$, for particular terminal points $q_1 = (x_1, y_1, \theta_1) \in \mathbb{R}^2 \times S^1$. Where applicable, we interpret the optimal trajectories in terms of the corresponding optimal motion of a car in the plane (x, y).

For generic terminal points, was developed a software in computer system Mathematica [116] for numerical evaluation of solutions to the problem.

4.2.11.1   The Case $x_1 \neq 0$, $y_1 = 0$, $\theta_1 = 0$

In this case $\lambda \in C_5$, and the optimal trajectory is

$$x_t = t \, \text{sgn} \, x_1, \quad y_t = 0, \quad \theta_t = 0, \quad t \in [0, t_1], \quad t_1 = |x_1|,$$

the car moves uniformly forward or backward along a line segment.

4.2.11.2   The Case $x_1 = 0$, $y_1 = 0$, $|\theta_1| \in (0, \pi)$

We have $\lambda \in C_4$, and the optimal solution is

$$x_t = 0, \qquad y_t = 0, \qquad \theta_t = t \operatorname{sgn} \theta_1, \qquad t \in [0, t_1], \qquad t_1 = |\theta_1|,$$

the car rotates uniformly around itself by the angle θ_1.

4.2.11.3   The Case $x_1 = 0$, $y_1 = 0$, $\theta_1 = \pi$

We have $\lambda \in C_4$, and there are two optimal solutions:

$$x_t = 0, \qquad y_t = 0, \qquad \theta_t = \pm t, \qquad t \in [0, t_1], \qquad t_1 = \pi,$$

the car rotates uniformly around itself clockwise or counterclockwise by the angle π.

4.2.11.4   The Case $x_1 \neq 0$, $y_1 = 0$, $\theta_1 = \pi$

There are two optimal solutions:

$$x_t = (\operatorname{sgn} x_1)/k(t + E(k) - E(K + t, k)), \qquad y_t = (s/k)\left(\sqrt{1 - k^2} - \operatorname{dn}(K + t, k)\right),$$

$$\theta_t = s \operatorname{sgn} x_1(\pi/2 - \operatorname{am}(K + t, k)), \qquad s = \pm 1, \qquad t \in [0, t_1], \qquad t_1 = 2K,$$

and $k \in (0, 1)$ is the root of the equation

$$(2/k)(K(k) - E(k)) = |x_1|.$$

4.2.11.5   The Case $x_1 = 0$, $y_1 \neq 0$, $\theta_1 = \pi$

There are two optimal solutions:

$$x_t = s(1 - \operatorname{dn}(t, k))/k, \qquad y_t = (\operatorname{sgn} y_1/k)(t - E(t, k)),$$

$$\theta_t = s \operatorname{sgn} y_1 \operatorname{am}(t, k), \qquad s = \pm 1, \qquad t \in [0, t_1], \qquad t_1 = 2K,$$

and $k \in (0, 1)$ is the root of the equation

$$(2/k)(K(k) - E(k)) = |y_1|.$$

4.2.11.6 The Case $x_1 = 0$, $y_1 \neq 0$, $\theta_1 = 0$

There are two optimal solutions given by formulas for (x_t, y_t, θ_t) for the case $\lambda \in C_2$ in Subsec. 3.3 [81] for the following values of parameters:

$$t \in [0, t_1], \quad t_1 = 2k p_1(k),$$

with the function $p_1(k)$ defined in Sect. 4.2.7,

$$s_2 = -\operatorname{sgn} y_1, \qquad \psi = \pm K(k) - p_1(k),$$

and $k \in (0, 1)$ is the root of the equation

$$2(p_1(k) - E(p_1(k), k)\sqrt{1 - k^2}/\operatorname{dn}(p_1(k), k)) = |y_1|.$$

4.2.11.7 The Case $(x_1, y_1) \neq 0$, $\theta_1 = \pi$

Introduce the polar coordinates $x_1 = \rho_1 \cos \chi_1$, $y_1 = \rho_1 \sin \chi_1$. There are two optimal solutions given by formulas for (x_t, y_t, θ_t) for the case $\lambda \in C_1$ in Subsec. 3.3 [81] for the following values of parameters:

$$t \in [0, t_1], \quad t_1 = 2K(k),$$

and $k \in (0, 1)$ is the root of the equation

$$2(p_1(k) - E(p_1(k), k)\sqrt{1 - k^2}/\operatorname{dn}(p_1(k), k)) = \rho_1,$$
$$s_1 = \pm 1, \qquad \varphi = s_1 F(\pi/2 - \chi_1, k).$$

In the cases $y_1 = 0$ and $x_1 = 0$ we get respectively the cases considered in Sects. 4.2.11.4 and 4.2.11.5.

4.2.11.8 The Case $y_1 \neq 0$, $\theta_1 = 0$

There is a unique optimal solution given by formulas for (x_t, y_t, θ_t) for the case $\lambda \in C_2$ in Subsec. 3.3 [81] for the following values of parameters:

$$s_2 = -\operatorname{sgn} y_1,$$

$k \in (0, 1)$ and $p \in (0, p_1(k)]$ are solutions to the system of equations

$$s(\text{sgn } y_1)2kf_1(p, k)/\,\text{dn}(p, k) = x_1, \qquad s = \pm 1,$$

$$2(p - \text{E}(p))\sqrt{1 - k^2}/\,\text{dn}(p, k) = |y_1|,$$

and

$$t \in [0, t_1], \qquad t_1 = 2kp, \qquad \psi = sK(k) - p.$$

4.2.12   Sub-Riemannian Spheres and Wavefronts

In this problem, sub-Riemannian spheres

$$S_R = \{q \in \text{SE}(2) \mid d(\text{Id}, q) = R\}$$

can be of four different topological types:

1. the zero radius sphere is a point
2. if $R \in (0, \pi)$, then S_R is homeomorphic to the 2-dimensional sphere S^2; see Fig. 4.24
3. for $R = \pi$ the sphere S_R is homeomorphic to the sphere S^2 with identified North pole N and South pole S: $S_\pi \cong S^2/\{N = S\}$; see Fig. 4.25
4. and if $R > \pi$, then S_R is homeomorphic to the 2-dimensional torus $\mathbb{T}^2$; see Fig. 4.26.

The spheres have singularities when they intersect the closure of the cut locus. Figure 4.27 shows self-intersections of the *wavefront*

$$W_R = \{\text{Exp}(\lambda, R) \mid \lambda \in C\}$$

for $R = \pi$.

Fig. 4.24 Sub-Riemannian sphere $S_{\pi/2} \subset \text{SE}(2)$

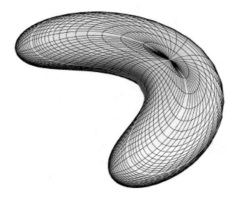

Fig. 4.25 Sub-Riemannian
sphere $S_\pi \subset SE(2)$

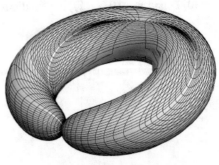

Fig. 4.26 Sub-Riemannian
sphere $S_{3\pi/2} \subset SE(2)$

Fig. 4.27 Wavefront
$W_\pi \subset SE(2)$ in cut-out

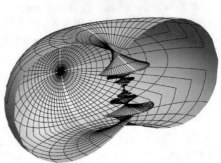

4.3   Euler's Elasticae

4.3.1   History of the Problem

A detailed account of the remarkable history of the problem on elasticae can be
found in [105, 106, 112, 113].

In 1691, J. Bernoulli [71] considered the problem on the form of a uniform
planar elastic bar bent by an external force. His hypothesis was that the bending
moment of the bar is equal to $\mathcal{B}/R$, where $\mathcal{B}$ is the "flexural rigidity" and R is the
curvature radius of the bent bar. For an elastic bar of unit excursion built vertically

Fig. 4.28 Rectangular
elastica

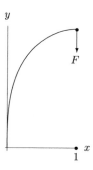

into a horizontal wall and bent by a load F sufficient to make its top horizontal
(*rectangular elastica*); see Fig. 4.28, J. Bernoulli obtained the ODEs

$$dy = \frac{x^2 \, dx}{\sqrt{1 - x^4}}, \quad ds = \frac{dx}{\sqrt{1 - x^4}}, \quad x \in [0, 1]$$

(where (x, y) is the elastic bar and s is its length parameter), integrated them in
series, and calculated precise upper and lower bounds for their value at the endpoint
$x = 1$.

In 1742, D. Bernoulli in his letter [70] to L. Euler wrote that the elastic energy of
the bent rod is proportional to the magnitude $J = \int ds/R^2$, where R is the curvature
radius of the rod, and suggested to find the elastic curves from the variational
principle $J \to \min$. At that time, Euler was writing his treatise on the calculus
of variations *Methodus inveniendi ...* [20] published in 1744, and he added an
appendix *De curvis elasticis*, where he applied the newly developed techniques
to the problem on elasticae. Euler considered a thin homogeneous elastic plate,
rectilinear in the natural (unstressed) state. For the profile of the plate, Euler stated
the following problem:

> "*That among all curves of the same length which not only pass*
>
> *through the points A and B, but are also tangent to given straight*
>
> *lines at these points, that curve be determined in which the value of* (4.49)
>
> $\int_A^B \dfrac{ds}{R^2}$ *be a minimum.*"

Euler wrote down the ODE known now as the Euler–Lagrange equation for the
corresponding problem of calculus of variations and reduced it to the equations

$$dy = \frac{(\alpha + \beta x + \gamma x^2) \, dx}{\sqrt{a^4 - (\alpha + \beta x + \gamma x^2)^2}}, \quad ds = \frac{a^2 \, dx}{\sqrt{a^4 - (\alpha + \beta x + \gamma x^2)^2}}, \quad (4.50)$$

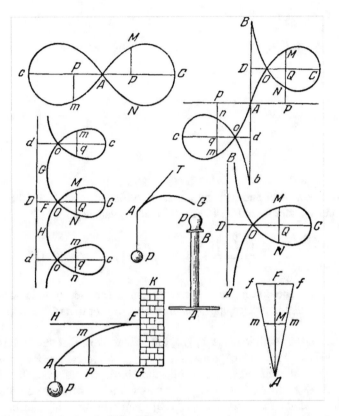

Fig. 4.29 Sketches of elasticae by Euler [20]

where α/a^2, β/a, and γ are real parameters expressed in terms of $\mathcal{B}$, the load of
the elastic rod, and its length. Euler studied the quadrature defined by the first of
Eqs. (4.50). In the modern terminology, he investigated the qualitative behaviour of
the elliptic functions that parametrize the elastic curves via the qualitative analysis of
the determining ODEs. Euler described all possible types of elasticae and indicated
the values of parameters for which these types are realized (see a copy of Euler's
original sketches in Fig. 4.29).

Euler divided all elastic curves into nine classes, they are plotted respectively as
follows:

1. a straight line, Fig. 4.38
2. Fig. 4.39
3. a rectangular elastica, Fig. 4.40
4. Fig. 4.41
5. a periodic elastica in the form of figure eight, Fig. 4.42
6. Fig. 4.43
7. a critical elastica with one loop, Fig. 4.44
8. Fig. 4.45
9. a circle, Fig. 4.46.

Following the tradition introduced by A.E.H. Love [106], the elastic curves with inflection points (classes 2–6) are said to be *inflectional*, the elastica of class 7 is said to be *critical*, and elasticae without inflection points of class 8 are said to be *noninflectional*.

Further, Euler established the magnitude of the force applied to the elastic plate that results in each type of elasticae. He indicated the experimental method for evaluation of the flexural rigidity of the elastic plate by its form under bending. Finally, he studied the problem of stability of a column modelled by the loaded rod whose lower end is constrained to remain vertical, by presenting it as an elastica of the class 2 close to the straight line (a sinusoid). After the work of Leonhard Euler, the elastic curves are called *Euler elasticae*.

The first explicit parametrization of Euler elasticae was performed by L. Saalschütz in 1880 [87].

In 1906, the future Nobel prize-winner Max Born defended his Ph.D. thesis called *Stability of elastic lines in the plane and the space* [74]. Born considered the problem on elasticae as a problem of calculus of variations and derived from the Euler–Lagrange equation that its solutions $(x(t), y(t))$ satisfy the ODEs

$$\dot{x} = \cos\theta, \quad \dot{y} = \sin\theta,$$

$$A\ddot{\theta} + R\sin(\theta - \gamma) = 0, \quad A, R, \gamma = \text{const}, \qquad (4.51)$$

and, therefore, the angle θ defining the slope of elasticae satisfies the equation of the *mathematical pendulum* (4.51). Further, Born studied stability of elasticae with fixed endpoints and fixed tangents at the endpoints. Born proved that an elastic arc without inflection points is stable. In this case, the angle θ is monotone and, therefore, it can be taken as a parameter along the elastica; Born showed that the second variation of the functional of the elastic energy $J = \dfrac{1}{2}\int \dot{\theta}^2 \, dt$ is positive. In the general case, Born wrote down the Jacobian that vanishes at conjugate points. Since the functions entering this Jacobian were too complicated, Born restricted himself to numerical investigation of conjugate points. He was the first to plot elasticae numerically and verify the theory against experiments on elastic rods; see photos from his thesis in Fig. 4.30. Moreover, Born studied stability of Euler elasticae with various other boundary conditions and obtained some results for elastic curves in $\mathbb{R}^3$.

In 1993, V. Jurdjevic [80] discovered that Euler elasticae appear in the plate-ball problem; see Sect. 1.1.1.6.

In the same year, R. Brockett and L. Dai [76] discovered that Euler elasticae are projections of sub-Riemannian geodesics on the Cartan group; see Example 2 in Sect. 4.5 and [63, 96].

Finally, in 2021 it was shown that during an optimal motion of the mobile robot considered in Sect. 4.2, the endpoint of the vector modelling the robot moves along a non-inflectional elastica or along a line [62].

Elasticae were considered in the approximation theory as nonlinear splines, in computer vision as a maximum likelihood reconstruction of occluded edges, their 3-dimensional analogues are used in the modelling of DNA minicircles, etc.

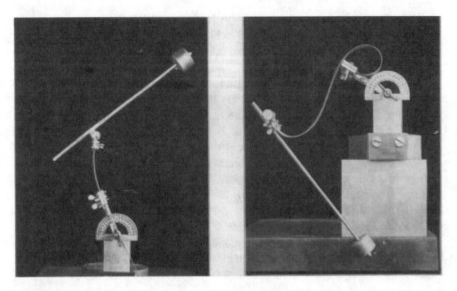

Fig. 4.30 Photos of experiments by Born with elasticae [74]

Euler elasticae and their various generalizations play an important role in modern mathematics, mechanics, and their applications. However, the initial variational problem as it was stated by Euler (4.49) is far from the complete solution. In this section we briefly present the main results known on this problem; see details in [89, 93, 94].

4.3.2 Problem Statement

We have the following optimal control problem; see Sect. 1.1.1.5:

$$\dot{x} = \cos\theta, \qquad q = (x, y, \theta) \in \mathbb{R}^2_{x,y} \times S^1_\theta = M, \qquad (4.52)$$

$$\dot{y} = \sin\theta, \qquad u \in \mathbb{R}, \qquad\qquad\qquad\qquad (4.53)$$

$$\dot{\theta} = u, \qquad\qquad\qquad\qquad\qquad\qquad (4.54)$$

$$q(0) = q_0 = (0, 0, 0), \qquad q(t_1) = q_1, \qquad (4.55)$$

$$t_1 \text{ is fixed}, \qquad\qquad\qquad\qquad\qquad (4.56)$$

$$J = \frac{1}{2} \int_0^{t_1} u^2 \, dt \to \min. \qquad\qquad (4.57)$$

Choosing an appropriate unit of length in the plane $\mathbb{R}^2_{x,y}$, we can assume that $t_1 = 1$.

Similarly to the sub-Riemannian problem on the group SE(2), problem (4.52)–(4.57) is left-invariant on the same group.

4.3.3 Existence of Solutions

As we showed in Sect. 2.3.2, $O_{q_0} = M$ and

$$\mathcal{A}_{q_0}(1) = \{q \in M \mid x^2 + y^2 < 1 \text{ or } (x, y, \theta) = (1, 0, 0)\}.$$

We suppose below that $q_1 \in \mathcal{A}_{q_0}(1)$. The set of control parameters $U = \mathbb{R}$ is noncompact, thus the Filippov theorem is not applicable. One can show (using general existence results of optimal control theory [108]) that optimal control exists; see [89].

4.3.4 The Pontryagin Maximum Principle

Denote the left-invariant frame on $M \cong SE(2)$:

$$f_1 = \cos\theta \frac{\partial}{\partial x} + \sin\theta \frac{\partial}{\partial y},$$

$$f_2 = \frac{\partial}{\partial \theta},$$

$$f_3 = \sin\theta \frac{\partial}{\partial x} + \sin\theta \frac{\partial}{\partial y},$$

and introduce the linear on fibers Hamiltonians—coordinates on $T_q^* M$:

$$h_i(\lambda) = \langle \lambda, f_i \rangle, \qquad i = 1, 2, 3.$$

Then the Hamiltonian of PMP reads

$$h_u^\nu(\lambda) = \langle \lambda, f_1 + u f_2 \rangle + \frac{\nu}{2} u^2 = h_1 + u h_2 + \frac{\nu}{2} u^2.$$

The corresponding Hamiltonian system of PMP has the form:

$$\dot{h}_1 = \{h_1 + u h_2, h_1\} = -u h_3,$$

$$\dot{h}_2 = \{h_1 + u h_2, h_2\} = h_3,$$

$$\dot{h}_3 = \{h_1 + u h_2, h_3\} = u h_1,$$

$$\dot{q} = f_1 + u f_2.$$

The maximality condition of PMP is

$$h_1 + uh_2 + \frac{v}{2}u^2 \to \max_{u \in \mathbb{R}},$$

and the nontriviality condition reads

$$(h_1, h_2, h_3, v) \neq 0.$$

4.3.4.1 Abnormal Case

Let $v = 0$. Then the maximality condition $h_1 + uh_2 \to \max_{u \in \mathbb{R}}$ yields $h_2 \equiv 0$, whence from the Hamiltonian system $h_3 \equiv 0$. Then from the nontriviality condition of PMP $h_1 \neq 0$. The Hamiltonian system yields $u \equiv 0$.

The abnormal extremal trajectory is $q(t) = e^{tf_1}(q_0)$; it is projected to the line $(x, y)(t) = (t, 0)$. It is optimal since in this case $J = 0 = \min$, this corresponds to the elastic rod without forces applied.

4.3.4.2 Normal Case

Let $v = -1$. The maximality condition $h_1 + uh_2 - \frac{u^2}{2} \to \max_{u \in \mathbb{R}}$ implies $u = h_2$, then the Hamiltonian system $\dot{\lambda} = \mathbf{H}(\lambda)$ with the maximized Hamiltonian $H = h_1 + \frac{h_2^2}{2}$ reads as follows:

$$\dot{h}_1 = -h_2 h_3, \tag{4.58}$$

$$\dot{h}_2 = h_3, \tag{4.59}$$

$$\dot{h}_3 = h_2 h_1, \tag{4.60}$$

$$\dot{q} = f_1 + h_2 f_2. \tag{4.61}$$

In addition to the Hamiltonian H, this system has an integral $F = h_1^2 + h_3^2$. Trajectories of the vertical subsystem (4.58)–(4.60) are obtained by intersection of the level surfaces of the integrals H and F; see Fig. 4.31 for the case $F > 0$. Different qualitative types of intersections of a circular cylinder $\{F = r > 0\}$ with the parabolic cylinders $H = \text{const}$ are shown in Figs. 4.32, 4.33, 4.34 and 4.35. In the case $F = 0$ we have $h_1 = h_3 = 0$, $h_2 \equiv \text{const}$.

Introduce the polar coordinates

$$h_1 = r \cos \alpha, \quad h_3 = r \sin \alpha,$$

Fig. 4.31 Intersection of
level surfaces of integrals H
and $F > 0$

Fig. 4.32 Intersection of
$\{F = r > 0\}$ and $\{H = -r\}$

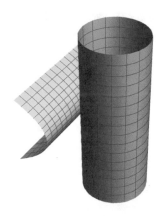

Fig. 4.33 Intersection of
$\{F = r > 0\}$ and
$\{H = \text{const} \in (-r, r)\}$

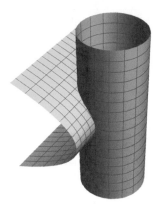

Fig. 4.34 Intersection of
$\{F = r > 0\}$ and $\{H = r\}$

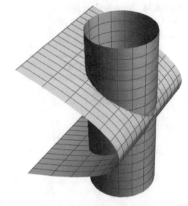

Fig. 4.35 Intersection of
$\{F = r > 0\}$ and $\{H > r\}$

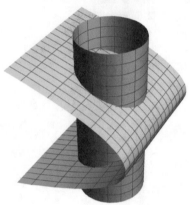

then the vertical subsystem (4.58)–(4.60) reads as follows:

$$\dot{\alpha} = h_2,$$

$$\dot{h}_2 = r \sin \alpha,$$

$$\dot{r} = 0.$$

Denote $\beta = \alpha + \pi$, $c = h_2$, then we get the equation of pendulum:

$$\dot{\beta} = c, \qquad c \in \mathbb{R}, \tag{4.62}$$

$$\dot{c} = -r \sin \beta, \qquad \beta \in S^1, \tag{4.63}$$

see Figs. 4.36 and 4.37. The pendulum is *Kirchhoff's kinetic analogue* of Euler elasticae; see [26, 106].

Fig. 4.36 Mathematical
pendulum

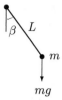

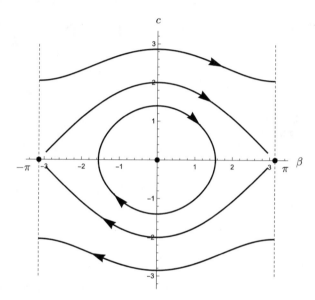

Fig. 4.37 Phase portrait of pendulum (4.62) and (4.63)

The equation of pendulum has the full energy integral $E = H = h_1 + \frac{h_2^2}{2} = \frac{c^2}{2} - r\cos\beta$. The ODEs for the horizontal variables are as follows:

$$\dot{x} = \cos\theta, \tag{4.64}$$

$$\dot{y} = \sin\theta, \tag{4.65}$$

$$\dot{\theta} = c = \dot{\beta}, \quad \text{thus } \theta = \beta - \beta_0. \tag{4.66}$$

The shape of Euler elasticae $(x(t), y(t))$ is determined by values of the energy integral $E \in [-r, +\infty)$ and the corresponding motion of pendulum (4.62) and (4.63).

- If $E = -r < 0$, then the pendulum stays at the stable equilibrium $(\beta, c) = (0, 0)$, and the elastic curve is a straight line; see Fig. 4.38.

Fig. 4.38 Elastica-line :
$E = \pm r, r > 0, c = 0$

Fig. 4.39 Inflectional
elastica: $E \in (-r, r), r > 0$,
$k \in (0, 1/\sqrt{2})$

Fig. 4.40 Rectangular
inflectional elastica:
$E \in (-r, r), r > 0$,
$k = 1/\sqrt{2}$

Fig. 4.41 Inflectional
elastica: $E \in (-r, r), r > 0$,
$k \in (1/\sqrt{2}, k_0)$

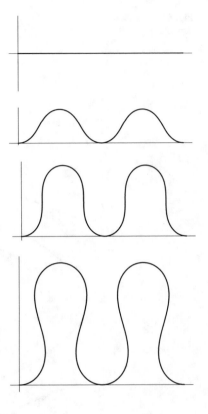

- If $E \in (-r, r), r > 0$, then the pendulum oscillates, and Euler elasticae have inflection points; see Figs. 4.39, 4.40, 4.41, 4.42 and 4.43.
- If $E = r > 0$, then the pendulum either stays at the unstable equilibrium $(\beta, c) = (\pi, 0)$, or tends to it for an infinite time; correspondingly Euler elasticae are either a straight line or a critical curve without inflection points and with one loop; see Fig. 4.44.
- If $E > r > 0$, then the pendulum rotates in one or another direction, and elastic curves have no inflection points; see Fig. 4.45.
- Finally, if $r = 0$, then the pendulum either rotates uniformly or stays fixed (in the absence of gravity); the elastic curves are respectively either circles (see Fig. 4.46) or straight lines.

In Figs. 4.39, 4.40, 4.41, 4.42 and 4.43 we use the parameter $k = \sqrt{(E + r)/(2r)} \in (0, 1)$.

In Fig. 4.47 we present all Euler elasticae as a one-parameter family connecting the line with the circle, and in Fig. 4.29 we show sketches of elasticae from Euler's book [20].

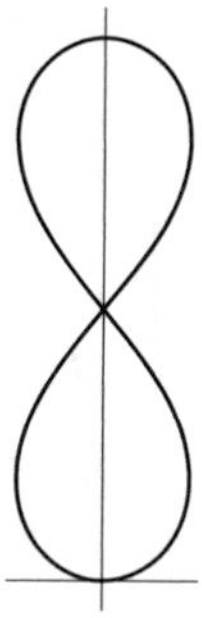

Fig. 4.42 Periodic
inflectional elastica:
$E \in (-r, r), r > 0$,
$k = k_0 \approx 0.909$

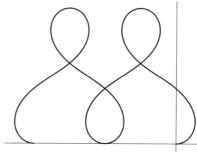

Fig. 4.43 Inflectional
elastica: $E \in (-r, r), r > 0$,
$k \in (k_0, 1)$

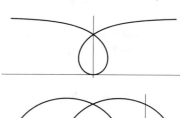

Fig. 4.44 Critical elastica:
$E = r > 0, \beta \neq \pi$

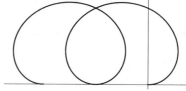

Fig. 4.45 Noninflectional
elastica: $E > r > 0$

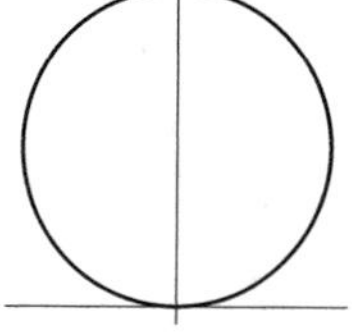

Fig. 4.46 Elastica-circle:
$r = 0, c \neq 0$

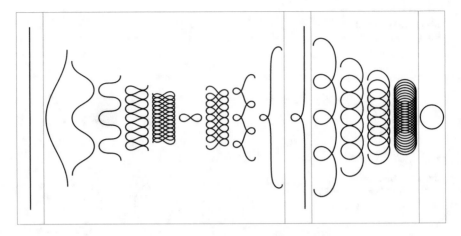

Fig. 4.47 One-parameter family of shapes of Euler elasticae

4.3.5 Exponential Mapping and Its Symmetries

The equations of pendulum (4.62) and (4.63) are integrable in Jacobi's elliptic functions [104, 115]; see Sect. A.2. As a consequence, ODEs (4.64)–(4.66) were also integrated in these functions; see [89].

The *exponential mapping* for a time $t_1 > 0$ in the problem on elasticae is defined as

$$\text{Exp}_{t_1} \ : \ N = T_{q_0}^* M \to M, \qquad \lambda \mapsto q(t_1) = \pi \circ e^{t_1 \mathbf{H}}(\lambda).$$

In view of the previous paragraph, this mapping is parametrized by Jacobi's elliptic functions.

The preimage of the exponential mapping N admits a decomposition into invariant subsets of the normal Hamiltonian system of PMP (4.58)–(4.61) by critical sets of the energy E:

$$N = \bigsqcup_{i=1}^{7} N_i,$$

$$N_1 = \{\lambda \in N \mid r \neq 0, \ E \in (-r, r)\},$$

$$N_2 = \{\lambda \in N \mid r \neq 0, \ E \in (r, +\infty)\},$$

$$N_3 = \{\lambda \in N \mid r \neq 0, \ E = r, \ \gamma \neq \pi\},$$

$$N_4 = \{\lambda \in N \mid r \neq 0, \ E = -r\},$$

$$N_5 = \{\lambda \in N \mid r \neq 0, \ E = r, \ \gamma = \pi\},$$

$$N_6 = \{\lambda \in N \mid r = 0, \ c \neq 0\},$$
$$N_7 = \{\lambda \in N \mid r = c = 0\}.$$

The domains N_1, N_2, and N_3 correspond to inflectional, non-inflectional, and to critical elasticae respectively, N_6 to circles, and $N_4 \cup N_5 \cup N_7$ to lines.

We will further use the following integral of the normal Hamiltonian system of PMP (4.58)–(4.61):

$$\lambda \in N_1 \quad \Rightarrow \quad k = \sqrt{\frac{E+r}{2r}} \in (0, 1),$$

$$\lambda \in N_2 \quad \Rightarrow \quad k = \sqrt{\frac{2r}{E+r}} \in (0, 1),$$

k is the modulus of Jacobi's functions that parametrize extremals.

Periodic motions of pendulum (4.62) and (4.63) have the period

$$T = \begin{cases} \frac{4}{\sqrt{r}} K(k), & \lambda \in N_1, \\ \frac{2}{\sqrt{r}} k K(k), & \lambda \in N_2, \\ \frac{2\pi}{|c|}, & \lambda \in N_6. \end{cases} \tag{4.67}$$

The phase portrait of pendulum (4.62) is preserved by the group of symmetries G generated by the reflection ε^1 in the axis β, the reflection ε^2 in the axis c, and the reflection ε^3 in the origin $(\beta, c) = (0, 0)$:

$$G = \{\mathrm{Id}, \varepsilon^1, \varepsilon^2, \varepsilon^3\} \simeq \mathbb{Z}_2 \times \mathbb{Z}_2.$$

These symmetries are naturally extended to the preimage N and image M of the exponential mapping Exp_t. If $\lambda = (\beta, c, r) \in N$, then

$$\varepsilon^i(\lambda) := \lambda^i = (\beta^i, c^i, r) \in N,$$

where

$$(\beta^1, c^1) = (\beta_t, -c_t),$$
$$(\beta^2, c^2) = (-\beta_t, c_t),$$
$$(\beta^3, c^3) = (-\beta, -c).$$

If $q = (x, y, \theta) \in M$, then $\varepsilon^i(q) := (x^i, y^i, \theta^i) \in M$, where

$$(x^1, y^1, \theta^1) = (x \cos\theta + y \sin\theta, -x \sin\theta + y \cos\theta, -\theta),$$
$$(x^2, y^2, \theta^2) = (x \cos\theta + y \sin\theta, x \sin\theta - y \cos\theta, \theta),$$
$$(x^3, y^3, \theta^3) = (x, -y, -\theta).$$

Proposition 4.4 *The group* $G = \{\mathrm{Id}, \varepsilon^1, \varepsilon^2, \varepsilon^3\}$ *consists of symmetries of the exponential mapping.*

Theorem 4.4 *Given an extremal trajectory* $q(t) = \mathrm{Exp}_t(\lambda)$, $\lambda \in N$, $t \geq 0$, *the first Maxwell time along the trajectory corresponding to the group of symmetries* G *is expressed as follows:*

$$\lambda \in N_1 \quad \Rightarrow \quad t^G_{\mathrm{Max}}(\lambda) = \frac{2}{\sqrt{r}} p_1(k) \in \left(\frac{T}{2}, T\right),$$

$$p_1(k) = \min(2K(k), p_z^1(k)) = \begin{cases} 2K(k), & k \in (0, k_0], \\ p_z^1(k), & k \in (k_0, 1), \end{cases} \qquad (4.68)$$

$$\lambda \in N_2 \quad \Rightarrow \quad t^G_{\mathrm{Max}}(\lambda) = \frac{2}{\sqrt{r}} kK(k) = T,$$

$$\lambda \in N_6 \quad \Rightarrow \quad t^G_{\mathrm{Max}}(\lambda) = \frac{2\pi}{|c|} = T,$$

$$\lambda \in N_3 \cup N_4 \cup N_5 \cup N_7 \quad \Rightarrow \quad t^G_{\mathrm{Max}}(\lambda) = +\infty.$$

Here $p = p_z^1(k) \in (K, 3K)$ *is the first positive root of the function* $f_1(p, k) = \mathrm{dn}\, p \,\mathrm{sn}\, p + (p - 2\,\mathrm{E}(p))\,\mathrm{cn}\, p = 0$, *and* $k_0 \approx 0.909$ *is the root of the equation* $2\mathrm{E}(k) - K(k) = 0$.

Proof See [89].   □

4.3.6   Bounds of Conjugate Time

For Euler elasticae the question of local optimality is important from the applied point of view since it means stability of the elastic rod w.r.t. its small perturbations with fixed endpoints and tangents at the endpoints. From the theoretical point of view, an answer to this question is essential as a step in the study of global optimality of elasticae. The results of this subsection were proved in [93].

Theorem 4.5 *Let* $\lambda \in N_1$. *Then the first conjugate time* $t^1_{\text{conj}}(\lambda)$ *on the trajectory* $\text{Exp}_t(\lambda)$ *belongs to the segment with the endpoints* $\dfrac{4K(k)}{\sqrt{r}}$ *and* $\dfrac{2p_1(k)}{\sqrt{r}}$, *namely:*

(1) $k \in (0, k_0) \quad \Rightarrow \quad t^1_{\text{conj}}(\lambda) \in \left[\dfrac{4K(k)}{\sqrt{r}}, \dfrac{2p^1_1(k)}{\sqrt{r}} \right],$

(2) $k = k_0 \quad \Rightarrow \quad t^1_{\text{conj}}(\lambda) = \dfrac{4K(k)}{\sqrt{r}} = \dfrac{2p^1_1(k)}{\sqrt{r}},$

(3) $k \in (k_0, 1) \quad \Rightarrow \quad t^1_{\text{conj}}(\lambda) \in \left[\dfrac{2p^1_1(k)}{\sqrt{r}}, \dfrac{4K(k)}{\sqrt{r}} \right],$

where the function $p_1(k)$ *is defined in* (4.68).

Corollary 4.4 *Let* $\lambda \in N_1$, $t_1 > 0$, *and let*

$$\Gamma = \{(x_t, y_t) \mid t \in [0, t_1]\}, \qquad q(t) = (x_t, y_t, \theta_t) = \text{Exp}_t(\lambda), \qquad (4.69)$$

be the corresponding elastic arc.

(1) *If the arc* Γ *does not contain inflection points, then it is stable.*
(2) *If* $k \in (0, k_0]$ *and the arc* Γ *contains exactly one inflection point, then it is stable.*
(3) *If the arc* Γ *contains not less than three inflection points in its interior, then it is unstable.*

Consider arcs of inflectional elasticae (4.69) centred at a vertex, i.e., let the curvature of the elastica attain a local extremum at its midpoint $(x_{t_1/2}, y_{t_1/2})$.

Denote $t^1_1 = \frac{2}{\sqrt{r}} p_1(k)$, where the function $p_1(k)$ is defined in (4.68).

Theorem 4.6 *Let an inflectional elastica* Γ *be centred at a vertex.*

(1) *If* $t < t^1_1$, *then the elastica* Γ *is stable.*
(2) *If* $t = t^1_1$, *then the endpoint of the elastica* Γ *is the first conjugate point.*
(3) *If* $t > t^1_1$, *then the elastica* Γ *is unstable.*

Now consider arcs of inflectional elasticae (4.69) centred at an inflection point, i.e., let an elastica have zero curvature at its midpoint $(x_{t_1/2}, y_{t_1/2})$.

Theorem 4.7 *Let an elastica* Γ *be centred at an inflection point. Let also* $k \in (0, k_0]$.

(1) *If* $t < T$, *then the elastica* Γ *is stable.*
(2) *If* $t = T$, *then the endpoint of the elastica* Γ *is the first conjugate point.*
(3) *If* $t > T$, *then the elastica* Γ *is unstable.*

Here T *is the period of elasticae; see* (4.67).

Theorem 4.8 *Let* $\lambda \in N_2 \cup N_3 \cup N_6$. *Then the extremal trajectory* $q(t) = \text{Exp}_t(\lambda)$ *does not have conjugate points for* $t > 0$.

Summing up: if an elastic arc does not have inflection points, then it is stable (this result was first obtained by M. Born [74]); if it has not less than three inflection points inside it, then it is unstable (thus one cannot hold an elastic rod with three inflection points inside it). If there are two or one inflection points, then the elastica can be stable or unstable.

4.3.7 Diffeomorphic Structure of the Exponential Mapping

Let $t_1 = 1$. Denote $\mathrm{Exp} = \mathrm{Exp}_1$ and

$$\mathcal{A} = \mathcal{A}_1(q_0) = \{(x, y, \theta) \in M \mid x^2 + y^2 < 1 \text{ or } (x, y, \theta) = (1, 0, 0)\}.$$

Consider the following subset of $\mathcal{A}$, which does not contain fixed points of the reflections $\varepsilon^1, \varepsilon^2$:

$$\tilde{M} = \{q \in \mathcal{A} \mid \varepsilon^i(q) \neq q, \ i = 1, 2\} = \left\{q \in \mathcal{A} \mid \sin\frac{\theta}{2} P(q) \neq 0\right\},$$

$$P(q) = x \sin\frac{\theta}{2} - y \cos\frac{\theta}{2},$$

and its decomposition into connected components

$$\tilde{M} = M_+ \sqcup M_-,$$

$$M_\pm = \{q \in M \mid \theta \in (0, 2\pi), \ x^2 + y^2 < 1, \ \mathrm{sgn}\, P(q) = \pm 1\}.$$

Consider also an open dense subset in the space of all potentially optimal extremal trajectories:

$$\tilde{N} = \{\lambda \in \cup_{i=1}^3 N_i \mid t_1 < t_{\mathrm{Max}}^G(\lambda), \ c_{t_1/2} \sin \beta_{t_1/2} \neq 0\},$$

and its connected components

$$\tilde{N} = \bigsqcup_{i=1}^4 D_i,$$

$$D_1 = \{\lambda \in \cup_{i=1}^3 N_i \mid t_1 < t_{\mathrm{Max}}^G(\lambda), \ c_{t_1/2} > 0, \ \sin \beta_{t_1/2} > 0\},$$

$$D_2 = \{\lambda \in \cup_{i=1}^3 N_i \mid t_1 < t_{\mathrm{Max}}^G(\lambda), \ c_{t_1/2} < 0, \ \sin \beta_{t_1/2} > 0\},$$

$$D_3 = \{\lambda \in \cup_{i=1}^3 N_i \mid t_1 < t_{\mathrm{Max}}^G(\lambda), \ c_{t_1/2} < 0, \ \sin \beta_{t_1/2} < 0\},$$

$$D_4 = \{\lambda \in \cup_{i=1}^3 N_i \mid t_1 < t_{\mathrm{Max}}^G(\lambda), \ c_{t_1/2} > 0, \ \sin \beta_{t_1/2} < 0\}.$$

Theorem 4.9 *The following mappings are diffeomorphisms:*

$$\mathrm{Exp} : D_1 \to M_+, \quad \mathrm{Exp} : D_2 \to M_-, \quad \mathrm{Exp} : D_3 \to M_+, \quad \mathrm{Exp} : D_4 \to M_-.$$

4.3.8 Optimal Trajectories for Various Boundary Conditions

4.3.8.1 Generic Boundary Conditions

If $q_1 \in M_+$, then there exists a unique pair $(\lambda_1, \lambda_3) \in D_1 \times D_3$ for which $\mathrm{Exp}(\lambda_1) = \mathrm{Exp}(\lambda_3) = q_1$, see Theorem 4.9. The optimal trajectory is among the trajectories $q^1(t) = \mathrm{Exp}_t(\lambda_1)$ and $q^3(t) = \mathrm{Exp}_t(\lambda_3)$, $t \in [0, 1]$. In order to find the optimal trajectory, one should take the trajectory for which the cost functional $J[q^i(\cdot)] = \frac{1}{2} \int_0^1 (c_t^i)^2 \, dt$ takes the minimum value. If $J[q^1(\cdot)] = J[q^3(\cdot)]$, then both trajectories are optimal.

If $q_1 \in M_-$, then the optimal trajectory is chosen similarly among the trajectories corresponding to the covectors $\lambda_2 \in D_2$ and $\lambda_4 \in D_4$ for which $\mathrm{Exp}(\lambda_2) = \mathrm{Exp}(\lambda_4) = q_1$.

4.3.8.2 The Case $(x_1, y_1, \theta_1) = (1, 0, 0)$

The optimal elastica is the line segment $(x, y)(t) = (t, 0)$, $t \in [0, 1]$.

4.3.8.3 The Case $x_1 > 0$, $y_1 = 0$, $\theta_1 = \pi$

In this case $q_1 \in M_+$ and the equation $\mathrm{Exp}(\lambda) = q_1$, $\lambda \in \widetilde{M}$, has two roots $\lambda_1 \in D_1$ and $\lambda_3 \in D_3$. The trajectories $q^1(t) = \mathrm{Exp}_t(\lambda_1)$ and $q^3(t) = \mathrm{Exp}_t(\lambda_3)$ have the same value of the functional J, thus are optimal. The corresponding optimal inflectional elasticae are symmetric w.r.t. the x axis; see Fig. 4.48.

Fig. 4.48 Optimal elasticae
for $x_1 > 0$, $y_1 = 0$, $\theta_1 = \pi$

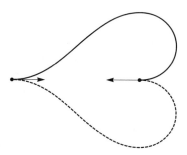

Fig. 4.49 Optimal elasticae
for $x_1 < 0$, $y_1 = 0$, $\theta_1 = \pi$

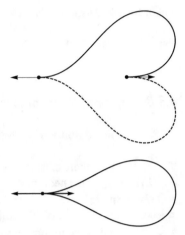

Fig. 4.50 Optimal
elastica-"drop" for $x_1 = 0$,
$y_1 = 0$, $\theta_1 = \pi$

4.3.8.4   The Case $x_1 < 0$, $y_1 = 0$, $\theta_1 = \pi$

This case is similar to the previous one; see Fig. 4.49.

4.3.8.5   The Case $x_1 = 0$, $y_1 = 0$, $\theta_1 = \pi$

There are two optimal elasticae defined by a covector $\lambda \in N_1$. They follow the "drop" given in Fig. 4.50 in two opposite directions.

4.3.8.6   The Case $x_1 = 0$, $y_1 = 0$, $\theta_1 = 0$

There are two optimal elasticae—circles symmetric in the x axis.

4.3.8.7   The Case $x_1 > 0$, $y_1 = 0$, $\theta_1 = 0$

There are two or four optimal elasticae; there exists $x_* \in (0.4,\ 0.5)$ such that:

- if $x_1 \in (0, x_*)$, then there are two optimal non-inflectional elasticae; see Fig. 4.51
- if $x_1 \in (x_*, 1)$, then there are two optimal inflectional elasticae; see Fig. 4.52
- if $x_1 = x_*$, then there are four optimal elasticae (two inflectional and two non-inflectional ones); see Fig. 4.53.

4.3.8.8   The Case $x_1 < 0$, $y_1 = 0$, $\theta_1 = 0$

There exist two optimal non-inflectional elasticae; see Fig. 4.54.

Fig. 4.51 Optimal elasticae
for $x_1 > 0$, $y_1 = 0$, $\theta_1 = 0$,
$x_1 \in (0, x_*)$

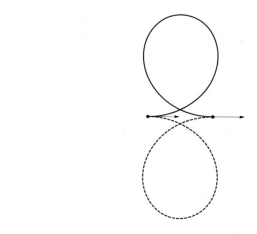

Fig. 4.52 Optimal elasticae
for $x_1 > 0$, $y_1 = 0$, $\theta_1 = 0$,
$x_1 \in (x_*, 1)$

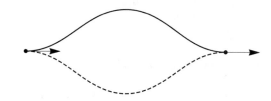

Fig. 4.53 Optimal elasticae
for $x_1 > 0$, $y_1 = 0$, $\theta_1 = 0$,
$x_1 = x_*$

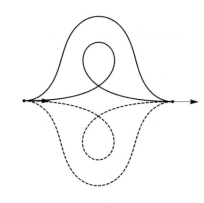

Fig. 4.54 Optimal elasticae
for $x_1 < 0$, $y_1 = 0$, $\theta_1 = 0$

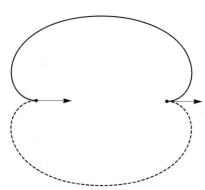

Summing up the results of this section: although the problem on elasticae was first considered in detail by Euler in 1742, the optimal synthesis is still unknown.

4.4   The Sub-Riemannian Problem on the Engel Group

4.4.1   Problem Statement

4.4.1.1   Geometric Statement

Consider the following generalization of Dido's problem. Let us have points a_0, $a_1 \in \mathbb{R}^2$, connected by a curve $\gamma_0 \subset \mathbb{R}^2$. Let us also have a number $S \in \mathbb{R}$ and a line $\ell \subset \mathbb{R}^2$. We should connect the points a_0, a_1 by the shortest curve $\gamma \subset \mathbb{R}^2$ so that γ_0 and γ bound in the plane a domain of the algebraic area S and a center of mass belonging to the line ℓ.

4.4.1.2   Optimal Control Problem

Similarly to Dido's problem, this problem can be stated as the following optimal control problem:

$$\dot{q} = u_1 X_1(q) + u_2 X_2(q), \qquad q = (x, y, z, v) \in \mathbb{R}^4, \quad u = (u_1, u_2) \in \mathbb{R}^2,$$
$$\tag{4.70}$$

$$q(0) = q_0, \qquad q(t_1) = q_1, \tag{4.71}$$

$$l = \int_0^{t_1} \sqrt{u_1^2 + u_2^2}\, dt \to \min, \tag{4.72}$$

$$X_1 = \frac{\partial}{\partial x} - \frac{y}{2}\frac{\partial}{\partial z}, \qquad X_2 = \frac{\partial}{\partial y} + \frac{x}{2}\frac{\partial}{\partial z} + \frac{x^2 + y^2}{2}\frac{\partial}{\partial v}. \tag{4.73}$$

This is the sub-Riemannian problem on $\mathbb{R}^4$ defined by the vector fields X_1, X_2 as an orthonormal frame.

4.4.1.3   The Engel Algebra and the Engel Group

The *Engel algebra* is the Lie algebra $\mathfrak{g}$ with a basis $(X_1, \ldots, X_4)$ in which nonzero Lie brackets are

$$[X_1, X_2] = X_3, \quad [X_1, X_3] = X_4,$$

Fig. 4.55 Engel algebra

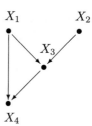

see Fig. 4.55.

The corresponding connected and simply connected Lie group M is called the *Engel group*. On the space $\mathbb{R}^4_{x,y,z,v}$, one can define a product rule

$$\begin{pmatrix} x_1 \\ y_1 \\ z_1 \\ v_1 \end{pmatrix} \cdot \begin{pmatrix} x_2 \\ y_2 \\ z_2 \\ v_2 \end{pmatrix} = \begin{pmatrix} x_1 + x_2 \\ y_1 + y_2 \\ z_1 + z_2 + (x_1 y_2 - x_2 y_1)/2 \\ v_1 + v_2 + y_1 y_2(y_1 + y_2)/2 + x_1 z_2 + x_1 y_2(x_1 + x_2)/2 \end{pmatrix},$$

that turns this space into the Engel group: $M \cong \mathbb{R}^4_{x,y,z,v}$, and the vector fields (4.73) into left-invariant vector fields on this Lie group. So problem (4.70)–(4.72) is a left-invariant sub-Riemannian problem on the Engel group. Thus we can assume that the initial point in (4.71) is the identity element of the Engel group: $q_0 = \mathrm{Id} = (0, 0, 0, 0)$.

One can show that all full-rank left-invariant sub-Riemannian problems with 2 controls on the Engel group are mutually isomorphic [88].

4.4.2 Geodesics

Existence of optimal controls in problem (4.70)–(4.72) follows from the Rashevskii–Chow and the Filippov theorems.

4.4.2.1 Pontryagin Maximum Principle

Let us pass from the length minimization problem (4.72) to the equivalent energy minimization problem

$$J = \frac{1}{2} \int_0^{t_1} (u_1^2 + u_2^2)\, dt \rightarrow \min. \tag{4.74}$$

Introduce the linear on fibers of T^*M Hamiltonians $h_i(\lambda) = \langle \lambda, X_i \rangle$, $i = 1, \ldots, 4$. Then PMP for problem (4.70), (4.71) and (4.74) takes the form:

$$\dot{h}_1 = -u_2 h_3,$$

$$\dot{h}_2 = u_1 h_3,$$

$$\dot{h}_3 = u_1 h_4,$$

$$\dot{h}_4 = 0,$$

$$\dot{q} = u_1 X_1 + u_2 X_2,$$

$$u_1 h_1 + u_2 h_2 + \frac{\nu}{2}(u_1^2 + u_2^2) \to \max_{(u_1, u_2) \in \mathbb{R}^2},$$

$$\nu \leq 0,$$

$$(h_1, \ldots, h_4, \nu) \neq 0.$$

4.4.2.2   Abnormal Extremals

Arclength parametrized abnormal minimizers are

$$h_1 = h_2 = h_3 = 0, \quad h_4 \equiv \text{const} \neq 0,$$

$$u_1 \equiv 0, \quad u_2 \equiv \pm 1,$$

$$x = z \equiv 0, \quad y = \pm t, \quad v = \pm \frac{t^3}{6}, \tag{4.75}$$

and are one-parameter subgroups $q(t) = e^{\pm t X_2}$.

4.4.2.3   Normal Extremals

Normal extremals are trajectories of the normal Hamiltonian system

$$\dot{\lambda} = \mathbf{H}(\lambda), \quad \lambda \in T^*M, \tag{4.76}$$

with the Hamiltonian $H = \frac{1}{2}(h_1^2 + h_2^2)$:

$$\dot{h}_1 = -h_2 h_3, \tag{4.77}$$

$$\dot{h}_2 = h_1 h_3, \tag{4.78}$$

$$\dot{h}_3 = h_1 h_4, \tag{4.79}$$

Fig. 4.56 Intersection of
level surfaces of integrals H,
$h_4 > 0$, and E

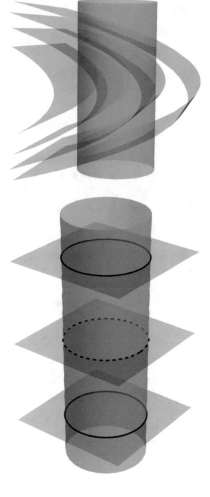

Fig. 4.57 Intersection of
level surfaces of integrals H,
$h_4 = 0$, and E

$$\dot{h}_4 = 0, \qquad\qquad\qquad\qquad\qquad (4.80)$$

$$\dot{q} = h_1 X_1 + h_2 X_2. \qquad\qquad\qquad\qquad (4.81)$$

Vertical subsystem (4.77)–(4.80) has integrals H, h_4, and $E = \frac{h_3^2}{2} - h_2 h_4$. Thus
its trajectories in the level surface $\{H = \frac{1}{2}\}$ are obtained by its intersection with
the parabolic cylinders $\{h_4 = h_4^0, \ E = E^0\}$ in the case $h_4^0 \neq 0$ (see Fig. 4.56),
and with the planes given by the same equations in the case $h_4^0 = 0$ (see Fig. 4.57).
The geometry of intersections of the circular cylinder $\{H = \frac{1}{2}\}$ with the parabolic
cylinders $\{h_4 = h_4^0, \ E = E^0\}$ in the case $h_4^0 \neq 0$ is the same as in Euler's elastic
problem; see Figs. 4.32, 4.33, 4.34 and 4.35.

Introduce coordinates (θ, c, α) on the level surface $\{H = 1/2\}$ as follows:

$$h_1 = -\sin\theta, \quad h_2 = \cos\theta, \quad h_3 = c, \quad h_4 = \alpha,$$

then the Hamiltonian system (4.76) takes the form

$$\dot\theta = c, \qquad \dot c = -\alpha\sin\theta, \qquad \dot\alpha = 0, \tag{4.82}$$

$$\dot q = -\sin\theta\, X_1 + \cos\theta\, X_2. \tag{4.83}$$

The vertical subsystem (4.82) is the equation of pendulum in the gravity field with the gravity acceleration $g = \alpha/L$, where L is the length of the pendulum. So in the case $\alpha > 0$ ($\alpha < 0$) the gravity is directed downwards (upwards) w.r.t. the axis from which the angle θ is measured, and in the case $\alpha = 0$ the pendulum moves in zero gravity.

Abnormal geodesics satisfy the normal Hamiltonian system (4.82) and (4.83) for $\theta = \pi + 2\pi n$, $c = 0$, thus they are *nonstrictly abnormal* (i.e., simultaneously normal and abnormal).

The family of normal extremals on the level surface $\{H = \frac{1}{2}\}$ is parametrized by initial points of the extremals belonging to the cylinder

$$C = T_{\mathrm{Id}}^* M \cap \left\{ H = \frac{1}{2} \right\}.$$

Consider a stratification of this cylinder into submanifolds corresponding to different types of trajectories of pendulum (4.82):

$$C = \bigsqcup_{i=1}^{7} C_i,$$

$$C_1 = \{\lambda \in C \mid \alpha \neq 0, \, E \in (-|\alpha|, |\alpha|)\},$$

$$C_2 = \{\lambda \in C \mid \alpha \neq 0, \, E \in (|\alpha|, +\infty)\},$$

$$C_3 = \{\lambda \in C \mid \alpha \neq 0, \, E = |\alpha|, c \neq 0\},$$

$$C_4 = \{\lambda \in C \mid \alpha \neq 0, \, E = -|\alpha|\},$$

$$C_5 = \{\lambda \in C \mid \alpha \neq 0, \, E = |\alpha|, c = 0\},$$

$$C_6 = \{\lambda \in C \mid \alpha = 0, \, c \neq 0\},$$

$$C_7 = \{\lambda \in C \mid \alpha = c = 0\}.$$

Periodic motions of pendulum (4.82) have the following periods:

$$\lambda \in C_1 \quad \Rightarrow \quad T = \frac{4K(k)}{\sqrt{|\alpha|}}, \quad k = \sqrt{\frac{E + |\alpha|}{2|\alpha|}},$$

$$\lambda \in C_2 \quad \Rightarrow \quad T = \frac{2kK(k)}{\sqrt{|\alpha|}}, \quad k = \sqrt{\frac{2|\alpha|}{E + |\alpha|}},$$

$$\lambda \in C_6 \quad \Rightarrow \quad T = \frac{2\pi}{|c|}.$$

In order to parametrize geodesics, on the strata C_1, C_2, C_3 straightening coordinates (φ, k, α) are introduced as in Sect. A.2.2 in which the vertical part (4.82) of the normal Hamiltonian system reads

$$\dot{\varphi} = 1, \qquad \dot{k} = 0, \qquad \dot{\alpha} = 0,$$

thus it has solutions

$$\varphi_t = \varphi + t, \qquad k = \text{const}, \qquad \alpha = \text{const}. \tag{4.84}$$

On this basis, a parametrization of normal extremals by Jacobi's elliptic functions was obtained.

Projections of geodesics to the plane (x, y) are Euler elasticae (see Sect. 4.3): inflectional ones for $\lambda \in C_1$, non-inflectional ones for $\lambda \in C_2$, critical ones for $\lambda \in C_3$, lines for $\lambda \in C_4 \cup C_5 \cup C_7$, and circles for $\lambda \in C_6$.

The family of all geodesics is parametrized by the exponential mapping

$$\text{Exp}: N = C \times \mathbb{R}_+ \to M,$$

$$\text{Exp}(\lambda, t) = \pi \circ e^{t\mathbf{H}}(\lambda) = q_t = (x_t, y_t, z_t, v_t).$$

4.4.3   Symmetries of the Exponential Mapping and Maxwell Time

Similarly to the problems considered in Sects. 4.2 and 4.3, the exponential mapping of the sub-Riemannian problem on the Engel group has a discrete group of symmetries

$$G = \{\text{Id}, \varepsilon^1, \dots, \varepsilon^7\} \cong \mathbb{Z}_2 \times \mathbb{Z}_2 \times \mathbb{Z}_2.$$

Theorem 4.10 *The first Maxwell time corresponding to the group of symmetries G is expressed as follows:*

$$\lambda \in C_1 \qquad\qquad \Rightarrow \quad t_{\text{Max}}^G(\lambda) = \frac{\min(2p_z^1(k), 4K(k))}{\sigma} \in \left(\frac{T}{2}, T\right),$$
(4.85)

$$\lambda \in C_2 \qquad\qquad \Rightarrow \quad t_{\text{Max}}^G(\lambda) = \frac{2kK(k)}{\sigma} = T, \qquad\qquad (4.86)$$

$$\lambda \in C_6 \qquad\qquad \Rightarrow \quad t_{\text{Max}}^G(\lambda) = \frac{2\pi}{|c|} = T, \qquad\qquad (4.87)$$

$$\lambda \in C_3 \cup C_4 \cup C_5 \cup C_7 \quad \Rightarrow \quad t_{\text{Max}}^G(\lambda) = +\infty, \qquad\qquad (4.88)$$

where $\sigma = \sqrt{|\alpha|}$, $p_z^1(k) \in \big(K(k), 3K(k)\big)$ *is the first positive root of the function* $f_z(p, k) = \text{dn } p \text{ sn } p + (p - 2E(p)) \text{ cn } p$, *and T is the period of pendulum (4.82).*

4.4.4   Lower Bound of the First Conjugate Time

Similarly to Sect. 4.2.8, one has the following bound.

Theorem 4.11 *For any* $\lambda \in C$

$$t_{\text{conj}}^1(\lambda) \geq t_{\text{Max}}^G(\lambda).$$

4.4.5   Diffeomorphic Structure of the Exponential Mapping

Consider a subset in the state space not containing fixed points of the symmetries ε^1 and ε^2:

$$\widetilde{M} = \{q \in M \mid \varepsilon^1(q) \neq q \neq \varepsilon^2(q)\} = \{q \in M \mid xz \neq 0\},$$

and its connected components:

$$M_1 = \{q \in M \mid x < 0, \ z > 0\},$$
$$M_2 = \{q \in M \mid x < 0, \ z < 0\},$$
$$M_3 = \{q \in M \mid x > 0, \ z < 0\},$$
$$M_4 = \{q \in M \mid x > 0, \ z > 0\}.$$

Consider also an open dense subset in the space of all potentially optimal geodesics:

$$\widetilde{N} = \left\{ (\lambda, t) \in N \mid t < t^G_{\text{Max}}(\lambda),\ c_{t/2} \sin \theta_{t/2} \neq 0 \right\},$$

and its connected components:

$$D_1 = \left\{ (\lambda, t) \in N \mid t < t^G_{\text{Max}}(\lambda),\ \sin \theta_{t/2} > 0,\ c_{t/2} > 0 \right\},$$

$$D_2 = \left\{ (\lambda, t) \in N \mid t < t^G_{\text{Max}}(\lambda),\ \sin \theta_{t/2} > 0,\ c_{t/2} < 0 \right\},$$

$$D_3 = \left\{ (\lambda, t) \in N \mid t < t^G_{\text{Max}}(\lambda),\ \sin \theta_{t/2} < 0,\ c_{t/2} < 0 \right\},$$

$$D_4 = \left\{ (\lambda, t) \in N \mid t < t^G_{\text{Max}}(\lambda),\ \sin \theta_{t/2} < 0,\ c_{t/2} > 0 \right\}.$$

Theorem 4.12 *The following mappings are diffeomorphisms:*

$$\text{Exp} : D_i \to M_i, \qquad i = 1, \ldots, 4,$$
$$\text{Exp} : \widetilde{N} \to \widetilde{M}.$$

Proof Apply the Hadamard theorem.   □

4.4.6   Cut Time and Cut Locus

Theorem 4.13 *For any $\lambda \in C$*

$$t_{\text{cut}}(\lambda) = t^G_{\text{Max}}(\lambda).$$

Theorem 4.14 *The cut locus is contained in the union of the coordinate subspaces $\{x = 0\}$ and $\{z = 0\}$. It is invariant with respect to the dilations*

$$\delta_s : (t, x, y, z, v) \mapsto (e^s t, e^s x, e^s y, e^{2s} z, e^{3s} v), \qquad s \in \mathbb{R},$$

and discrete symmetries:

$$\delta_s(\text{Cut}) = \text{Cut}, \qquad s \in \mathbb{R},$$
$$\varepsilon^i(\text{Cut}) = \text{Cut}, \qquad i = 1, \ldots, 7.$$

An explicit stratification of the cut locus was obtained in [64].

4.4.7   *Sphere*

Sub-Riemannian spheres are transformed one into another by left translations

$$L_q(S_R(q_0)) = S_R(qq_0), \qquad q, \, q_0 \in M, \quad R \geq 0,$$

and dilations

$$\delta_s(S_R(\mathrm{Id})) = S_{R'}(\mathrm{Id}), \qquad R' = e^s R, \quad s \in \mathbb{R}, \quad R \geq 0,$$

thus it is enough to study the unit sphere $S = S_1(\mathrm{Id})$.

The unit sphere is invariant w.r.t. reflections:

$$\varepsilon^i(S) = S, \quad i = 1, \ldots, 7.$$

Consider the intersection of the unit sphere with the invariant manifold of the main symmetries $\varepsilon^1, \varepsilon^2$:

$$\tilde{S} = \{q \in S \mid \varepsilon^1(q) = \varepsilon^2(q) = q\} = S \cap \{x = z = 0\},$$

see Fig. 4.58. Here

$$A_{\pm} = \tilde{S} \cap \{w = 0, \, \mathrm{sgn}\, y = \pm 1\},$$
$$C_{\pm} = \tilde{S} \cap \{y = 0, \, \mathrm{sgn}\, w = \pm 1\},$$
$$\gamma_1 = \tilde{S} \cap \{y < 0, \, w > 0\},$$
$$\gamma_2 = \tilde{S} \cap \{y > 0, \, w > 0\},$$
$$\gamma_3 = \tilde{S} \cap \{y > 0, \, w < 0\},$$
$$\gamma_4 = \tilde{S} \cap \{y < 0, \, w < 0\},$$
$$\tilde{S} = A_+ \sqcup A_- \sqcup C_+ \sqcup C_- \sqcup \left(\sqcup_{i=1}^4 \gamma_i \right) \tag{4.89}$$

and $w = v - y^3/6$.

The section $\tilde{S}$ is centrally symmetric via the reflection ε^4:

$$\varepsilon^4(\gamma_i) = \gamma_{i+2}, \qquad i = 1, 2,$$
$$\varepsilon^4(A_+) = A_-, \qquad \varepsilon^4(C_+) = C_-.$$

Points of the section $\tilde{S}$ are characterized as follows:

- $A_{\pm}$ belong to abnormal minimizers
- $C_{\pm}$ are conjugate points, Maxwell points, and cut points

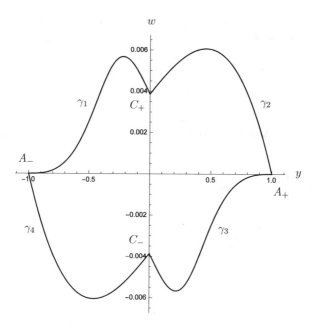

Fig. 4.58 Section of sphere $\widetilde{S} = S \cap \{x = z = 0\}$

- $q \in \gamma_i$ are Maxwell points and cut points.

Points of the section $\widetilde{S}$ have the following *multiplicity* μ (number of minimizers coming from Id to this point):

- $\mu(A_\pm) = 1$
- $\mu(C_\pm) = \mathfrak{c}$ (continuum $\cong S^1$)
- $q \in \gamma_i \quad \Rightarrow \quad \mu(q) = 2$.

There is a one-parameter family of minimizers coming to conjugate and cut points $C_\pm$. Unlike Dido's problem, these minimizers are not obtained one from another by a one-parameter group of symmetries. These minimizers project to the plane (x, y) to closed figure-of-eight elasticae.

Theorem 4.15 *The section $\widetilde{S}$ has the following regularity at its points:*

(1) *the curves γ_i are analytic and regular*
(2) $A_\pm$, $C_\pm$ *are singular points at which $\widetilde{S}$ is nonsmooth but Lipschitzian*
(3) $\overline{\gamma}_2 = \gamma_2 \cup \{C_+, A_+\}$ *is smooth of class C^∞*
(4) $\gamma_1 \cup \{C_+\}$ *is smooth of class C^∞*
(5) $\gamma_1 \cup \{A_-\}$ *is smooth of class C^1.*

A subset of a real-analytic manifold is called *analytic* if in some neighbourhood of each of its points it is defined by a finite system of real analytic equations. It is called *semi-analytic* if in some neighbourhood of each of its points it is defined by a finite system of real analytic equations and inequalities. Finally, it is called *subanalytic* if it can be obtained from semianalytic sets by a finite sequence of unions, intersections, and taking images of proper analytic mappings. In the two-dimensional plane, the systems of semianalytic and subanalytic sets coincide.

Theorem 4.16

(1) *The set $\widetilde{S} \setminus \{A_+, A_-\}$ is semianalytic.*
(2) *In a neighbourhood of the point A_- the curve γ_1 is a graph of a non-analytic function*

$$w = \frac{1}{6}Y^3 - 4Y^3 \exp(-2/Y)(1 + o(1)), \qquad Y = (y+1)/2 \to 0.$$

(3) *So the set $\widetilde{S}$ is not semianalytic, thus not subanalytic.*
(4) *Consequently, the sphere S is not subanalytic.*

4.4.8 Explicit Expressions of Sub-Riemannian Distance

For some points of the Engel group we know their distance to the identity element $d_0(q) = d(\mathrm{Id}, q)$:

- Abnormal minimizer $q(t) = e^{\pm t X_2}$, $x = z = w = 0$, $y = \pm t$, $t \geq 0$:

$$d_0(q(t)) = t.$$

- Central element of the Engel group $q(t) = e^{\pm t X_4}$, $x = y = z = 0$, $w = \pm t$, $t \geq 0$:

$$d_0(q(t)) = C \sqrt[3]{t},$$

$$C = \sqrt[3]{48 K^2(k_0)} \approx 6.37, \qquad K(k_0) - 2E(k_0) = 0, \quad k_0 \approx 0.909.$$

4.4.9 Metric Lines

A geodesic $q(t)$, $t \in \mathbb{R}$, of a sub-Riemannian structure is called a *metric line* if

$$d(q(t_1), q(t_2)) = |t_1 - t_2|, \qquad t_1, t_2 \in \mathbb{R}.$$

Theorem 4.17 *Only the following geodesics are metric lines on the Engel group:*

(1) *one-parameter subgroups tangent to the distribution:*

$$e^{t(u_1 X_1 + u_2 X_2)} = \mathrm{Exp}(\lambda, t), \quad t \in \mathbb{R}, \tag{4.90}$$

$$u_1 = -\sin\theta, \quad u_2 = \cos\theta, \quad \lambda = (\theta, c = 0, \alpha) \in C_4 \cup C_5,$$

(2) *critical geodesics:*

$$\mathrm{Exp}(\lambda, t), \quad \lambda \in C_3, \quad t \in \mathbb{R}. \tag{4.91}$$

Geodesics (4.90) project to the plane (x, y) into Euclidean lines, of which only the following ones are abnormal:

$$e^{t X_2} = \mathrm{Exp}(\lambda, t), \quad \lambda = (\theta = 0, c = 0, \alpha) \in C_4 \cup C_5.$$

Geodesics (4.91) project to the plane (x, y) into critical Euler elasticae (see Fig. 4.44), also called *Euler solitons* or *Euler kinks*.

The results of this section were obtained in [64, 97] and in papers cited therein.

4.5 Exercises

1. Prove that the product

$$(x_1, y_1, z_1) \cdot (x_2, y_2, z_2) = (x_1 + x_2, y_1 + y_2, z_1 + z_2 + (x_1 y_2 - x_2 y_1)/2),$$

$$(x_i, y_i, z_i) \in \mathbb{R}^3, \quad i = 1, 2,$$

   turns $\mathbb{R}^3$ into a Lie group called the *Heisenberg group*. Show that Dido's problem is left-invariant on this Lie group.
2. Show that in Dido's problem $d_0 \in C(\mathbb{R}^3)$, but $d_0 \notin C^1(q)$ for any $q = (0, 0, z)$, $z \in \mathbb{R}$.
3. Prove that the sub-Riemannian spheres in Dido's problem are semianalytic (thus subanalytic).
4. Prove that Euler's elastic problem is left-invariant on the Lie group SE(2).
5. Study the problem on the *Reeds–Shepp car:*

$$\dot{x} = u_1 \cos\theta, \quad q = (x, y, \theta) \in \mathbb{R}^2_{x,y} \times S^1_\theta = M,$$

$$\dot{y} = u_1 \sin\theta, \quad |u_1| = 1, \quad |u_2| \le 1,$$

$$\dot{\theta} = u_2,$$

$$q(0) = q_0, \quad q(t_1) = q_1,$$

$$t_1 \to \min.$$

(a) Prove that the problem is left-invariant on the Lie group SE(2).
(b) Show that the system is globally controllable.
(c) Show that abnormal trajectories are constant.
(d) Show that optimal trajectories are concatenations of arcs of unit circles and line segments.
(e) Study the existence and structure of optimal trajectories by [98].

6. Consider the following natural generalization of Dido's problem. Let us fix, in addition to two points (x_0, y_0), $(x_1, y_1) \in \mathbb{R}^2$, a curve $\gamma_0 \subset \mathbb{R}^2$ connecting them, a number M and a line $\ell \subset \mathbb{R}^2$. We should find the shortest curve $\gamma \subset \mathbb{R}^2$ that connects the points (x_0, y_0) and (x_1, y_1), such that the domain $D \subset \mathbb{R}^2$ bounded by the curves γ_0 and γ have the first moment of mass w.r.t. to the line ℓ equal to M.

   (a) Prove that this generalized Dido's problem can be stated as the following *sub-Riemannian problem in the flat Martinet case*:

   $$\dot{x} = u_1,$$
   $$\dot{y} = u_2,$$
   $$\dot{z} = \frac{1}{2}y^2 u_1,$$
   $$q = (x, y, z) \in \mathbb{R}^3, \qquad u = (u_1, u_2) \in \mathbb{R}^2,$$
   $$q(0) = q_0 = (0, 0, 0), \qquad q(t_1) = q_1,$$
   $$l = \int_0^{t_1} \sqrt{u_1^2 + u_2^2}\, dt \to \min.$$

   (b) Prove that the system is globally controllable.
   (c) Compute abnormal minimizers starting from q_0.
   (d) Prove that projections of sub-Riemannian minimizers to the plane (x, y) are inflectional Euler elasticae.
   (e) Parametrize normal geodesics by Jacobi's functions and elliptic integrals.
   (f) Assuming that geodesics lose optimality at the first intersection with the plane $\{y = 0\}$ (see [61]), describe the intersection of the sub-Riemannian sphere with this plane, similarly to Sect. 4.4.7. See the plots of the unit sphere and its intersection with the plane $\{y = 0\}$ in Figs. 4.59 and 4.60.

7. Consider another generalization of Dido's problem. Let us fix, in addition to two points (x_0, y_0), $(x_1, y_1) \in \mathbb{R}^2$, a curve $\gamma_0 \subset \mathbb{R}^2$ connecting them, and a number $S \in \mathbb{R}$, also a point $c \in \mathbb{R}^2$ in the plane. We should find the shortest curve $\gamma \subset \mathbb{R}^2$ that connects the points (x_0, y_0) and (x_1, y_1), such that the domain $D \subset \mathbb{R}^2$ bounded by the curves γ_0 and γ have the given algebraic area S and the centre of mass c.

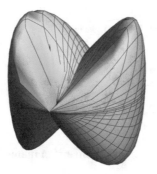

Fig. 4.59 Sub-Riemannian sphere S in the Martinet flat case

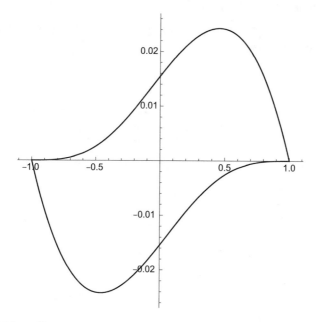

Fig. 4.60 $S \cap \{y = 0\}$

(a) Prove that this generalized Dido's problem can be stated as the following *sub-Riemannian problem on the Cartan group*:

$$\dot{x} = u_1,$$
$$\dot{y} = u_2,$$
$$\dot{z} = \frac{1}{2}(xu_2 - yu_1),$$
$$\dot{v} = \frac{1}{2}(x^2 + y^2)u_2,$$

$$\dot{w} = -\frac{1}{2}(x^2 + y^2)u_1,$$

$$q = (x, y, z, v, w) \in \mathbb{R}^5, \qquad u = (u_1, u_2) \in \mathbb{R}^2,$$

$$q(0) = q_0 = 0, \qquad q(t_1) = q_1,$$

$$l = \int_0^{t_1} \sqrt{u_1^2 + u_2^2}\, dt \to \min.$$

(b) Prove that projections of sub-Riemannian minimizers to the plane (x, y) are Euler elasticae.

(c) Compute the *abnormal set*, i.e., the set of all points in $\mathbb{R}^5$ filled by abnormal trajectories starting from q_0.

8. Prove that in the plate-ball problem, the sphere rolls optimally along Euler elasticae in the plane.

Chapter 5
Conclusion

Riding on the animal, he leisurely wends his way home:
Enveloped in the evening mist, how tunefully the flute vanishes away!
Singing a ditty, beating time, his heart is filled with a joy indescribable!
That he is now one of those who know, need it be told?

Pu-ming, "The Ten Oxherding Pictures" (cited by Suzuki [110])

Coming Home on the Ox's Back

We complete the book by recommendations for further reading.

© The Author(s), under exclusive license to Springer Nature Switzerland AG 2022
Yu. Sachkov, *Introduction to Geometric Control*, Springer Optimization
and Its Applications 192, https://doi.org/10.1007/978-3-031-02070-4_5

We recommend the following books for further study:

- M.I. Zelikin, Optimal control and calculus of variations [38]
- A.A. Agrachev, D. Barilari, U. Boscain, A comprehensive introduction to sub-Riemannian geometry from Hamiltonian viewpoint [2]
- A.A. Agrachev, Yu.L. Sachkov, Control theory from the geometric viewpoint [3]
- V. Jurdjevic, Geometric control theory [26]
- A.A. Ardentov, A.P. Mashtakov, A.V. Podobryaev, Yu.L. Sachkov, Symmetry method for optimal control problems on Lie groups [5].

In this book we used material of textbooks [2, 3, 27].

Appendix A
Elliptic Integrals, Functions and Equation of Pendulum

A.1 Elliptic Integrals and Functions

Standard references on elliptic integrals and functions are books [99, 104, 115]. We present here a minimal information on them required for the material of this book.

A.1.1 Elliptic Integrals in Legendre's Form

Incomplete elliptic integrals of the first kind:

$$F(\varphi, k) = \int_0^\varphi \frac{dt}{\sqrt{1 - k^2 \sin^2 t}},$$

of the second kind:

$$E(\varphi, k) = \int_0^\varphi \sqrt{1 - k^2 \sin^2 t}\, dt,$$

here and below the *elliptic modulus* $k \in (0, 1)$.

Complete elliptic integrals of the first and second kinds:

$$K(k) = F\left(\frac{\pi}{2}, k\right),$$

$$E(k) = E\left(\frac{\pi}{2}, k\right).$$

A.1.2 Jacobi's Functions

Jacobi's amplitude

$$\varphi = \operatorname{am}(u, k) \quad \Leftrightarrow \quad u = F(\varphi, k),$$

$$\operatorname{sn}(u, k) = \sin \operatorname{am}(u, k),$$

$$\operatorname{cn}(u, k) = \cos \operatorname{am}(u, k),$$

$$\operatorname{dn}(u, k) = \sqrt{1 - k^2 \operatorname{sn}^2(u, k)},$$

Jacobi's epsilon function

$$\mathrm{E}(u, k) = E(\operatorname{am} u, k).$$

In the notation of Jacobi's functions, the modulus k is often omitted.

A.1.3 Standard Formulae

Derivatives and integrals:

$$\operatorname{am}' u = \operatorname{dn} u,$$

$$\operatorname{sn}' u = \operatorname{cn} u \operatorname{dn} u,$$

$$\operatorname{cn}' u = -\operatorname{sn} u \operatorname{dn} u,$$

$$\operatorname{dn}' u = -k^2 \operatorname{sn} u \operatorname{cn} u,$$

$$\int_0^u \operatorname{dn}^2 t \, dt = \mathrm{E}(u).$$

Degeneration:

$$k \to +0 \quad \Rightarrow \quad \operatorname{sn} u \to \sin u, \quad \operatorname{cn} u \to \cos u, \quad \operatorname{dn} u \to 1, \quad \mathrm{E}(u) \to u,$$

$$k \to 1 - 0 \quad \Rightarrow \quad \operatorname{sn} u \to \tanh u, \quad \operatorname{cn} u, \operatorname{dn} u \to \frac{1}{\cosh u}, \quad \mathrm{E}(u) \to \tanh u.$$

Periods:

$$\operatorname{sn}(u + 4K) = \operatorname{sn} u, \qquad \operatorname{cn}(u + 4K) = \operatorname{cn} u, \qquad \operatorname{dn}(u + 2K) = \operatorname{dn} u.$$

A.2   Pendulum

In many left-invariant optimal control problems on Lie groups, the vertical subsystem of the Hamiltonian system of PMP mysteriously reduces to the equation of pendulum, thus they are all integrated in elliptic functions and integrals.

A.2.1   Equation of Pendulum and Its Solution

Consider a *mathematical pendulum*—a material point fixed at a weightless inextensible rod of length L, which can freely rotate in the vertical plane around the suspension point. Let θ denote the deviation angle of the pendulum from the bottom vertical position. Then the motion of the pendulum satisfies the equations

$$\dot{\theta} = c, \qquad \dot{c} = -r \sin \theta, \tag{A.1}$$

where $r = \dfrac{g}{L} > 0$ and g is the free gravity acceleration. The full energy of the pendulum (the first integral of equations (A.1)) is

$$E = \frac{c^2}{2} - r \cos \theta \in [-r, +\infty).$$

The character of motions of the pendulum is determined by the value of the energy E:

- if $E = -r$, then $(\theta, c) \equiv (0, 0)$, and the pendulum is at rest at the stable equilibrium
- if $E \in (-r, r)$, then the pendulum oscillates periodically around the stable equilibrium with the period $T = \frac{4}{\sqrt{r}} K(k)$, $k = \sqrt{\frac{E+r}{2r}} \in (0, 1)$, by the rule

$$\sin \frac{\theta}{2} = k \operatorname{sn}(\sqrt{r}\, t, k)$$

- if $E = r$, $c = 0$, then the pendulum is at rest at the unstable equilibrium $(\theta, c) \equiv (\pm\pi, 0)$
- if $E = r$, $c \neq 0$, then the pendulum performs a non-periodic motion along a separatrix and tends to the unstable equilibrium as $t \to \pm\infty$ by the rule

$$\sin \frac{\theta}{2} = \pm \tanh(\sqrt{r}\, t)$$

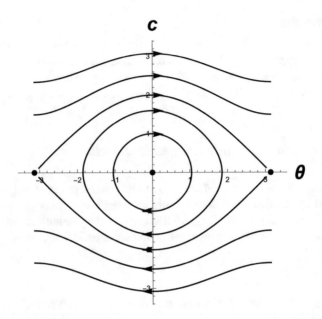

Fig. A.1 Phase portrait of pendulum (A.1)

- if $E > r$, then the pendulum rotates nonuniformly clockwise $(c < 0)$ or counterclockwise $(c > 0)$, with the period $T = \frac{2}{\sqrt{r}} k K(k)$, $k = \sqrt{\frac{2r}{E+r}} \in (0, 1)$, by the rule

$$\sin \frac{\theta}{2} = \pm \operatorname{sn}\left(\frac{\sqrt{r}}{k} t, k\right), \qquad \pm = \operatorname{sgn} c.$$

See the phase portrait of pendulum (A.1) for $r = 1$ in Fig. A.1.

We described above motions of pendulum (A.1) in the case $r = \frac{g}{L} > 0$. And if $r = 0$ (which can be interpreted as absence of gravity), then:

- for $c \neq 0$ the pendulum rotates uniformly clockwise $(c < 0)$ or counterclockwise $(c > 0)$
- for $c = 0$ the pendulum is at rest at a non-stable equilibrium.

The case $r < 0$ (gravity is directed upwards) reduces to the case $r > 0$ by the change of variables $(\theta, c, r) \mapsto (\theta + \pi, c, -r)$.

A.2.2 Straightening Coordinates

In the case $r > 0$ the phase cylinder of pendulum (A.1)

$$C = \{(\theta, c) \mid \theta \in S^1, \quad c \in \mathbb{R}\},$$

stratifies depending on the type of motion of the pendulum:

$$C = \bigsqcup_{i=1}^{5} C_i,$$

$$C_1 = \{(\theta, c) \in C \mid E \in (-r, r)\},$$
$$C_2 = \{(\theta, c) \in C \mid E > r\},$$
$$C_3 = \{(\theta, c) \in C \mid E = r, \quad c \neq 0\},$$
$$C_4 = \{(\theta, c) \in C \mid c = 0, \quad \theta = 0\},$$
$$C_5 = \{(\theta, c) \in C \mid c = 0, \quad \theta = \pi\}.$$

In the domains C_1, C_2, C_3 one can define coordinates (φ, k) that straighten the equation of pendulum: φ is the time of motion of the pendulum from a fixed point to a current point, and k is the first integral.
  If $(\theta, c) \in C_1$, then

$$k = \sqrt{\frac{E + r}{2r}} \in (0, 1), \qquad \sqrt{r}\varphi(\ \mathrm{mod}\ 4K(k)) \in [0, 4K(k)],$$

$$\sin\frac{\theta}{2} = k\,\mathrm{sn}(\sqrt{r}\varphi, k), \qquad \cos\frac{\theta}{2} = \mathrm{dn}(\sqrt{r}\varphi, k),$$

$$c = 2k\sqrt{r}\,\mathrm{cn}(\sqrt{r}\varphi, k).$$

If $(\theta, c) \in C_2$, then

$$k = \sqrt{\frac{2r}{E + r}} \in (0, 1), \qquad \sqrt{r}\varphi(\ \mathrm{mod}\ 2kK(k)) \in [0, 2kK(k)],$$

$$\sin\frac{\theta}{2} = \pm\,\mathrm{sn}\left(\frac{\sqrt{r}\varphi}{k}, k\right), \qquad \cos\frac{\theta}{2} = \mathrm{cn}\left(\frac{\sqrt{r}\varphi}{k}, k\right),$$

$$c = \pm 2\frac{\sqrt{r}}{k}\,\mathrm{dn}\left(\frac{\sqrt{r}\varphi}{k}, k\right), \qquad \pm = \mathrm{sgn}\,c.$$

If $(\theta, c) \in C_3$, then

$$k = 1, \qquad \varphi \in \mathbb{R},$$

$$\sin \frac{\theta}{2} = \pm \tanh(\sqrt{r}\varphi), \qquad \cos \frac{\theta}{2} = \frac{1}{\cosh(\sqrt{r}\varphi)},$$

$$c = \pm \frac{2\sqrt{r}}{\cosh(\sqrt{r}\varphi)}, \qquad \pm = \operatorname{sgn} c.$$

In the coordinates (φ, k) the equation of pendulum (A.1) straightens:

$$\dot{\varphi} = 1, \quad \dot{k} = 0,$$

thus it has solutions

$$\varphi_t = \varphi + t, \quad k \equiv \operatorname{const}.$$

These straightening coordinates were used in [89, 91] for explicit parametrization of extremal trajectories in the sub-Riemannian problem on the group SE(2) (Sect. 4.2), in Euler's elastic problem (Sect. 4.3), and in several other problems where appears the equation of pendulum, see [64, 77, 90, 93].

Bibliography and Further Reading

Books on Control Theory, Calculus of Variations, and Nonholonomic Geometry

1. A.A. Agrachev, *Some open problems*, Geometric Control Theory and Sub-Riemannian Geometry, pp. 1–13. Springer INdAM Series, vol. 5. Springer, 2014.
2. A. Agrachev, D. Barilari, U. Boscain, *A Comprehensive Introduction to sub-Riemannian Geometry from Hamiltonian viewpoint*, Cambridge University Press, 2019.
3. A.A. Agrachev, Yu.L. Sachkov, *Control theory from the geometric viewpoint*, Berlin Heidelberg New York Tokyo. Springer-Verlag, 2004.
4. V.M. Alekseev, V.M. Tikhomirov, and S.V. Fomin, *Optimal control*. New York: Consultants Bureau, 1987.
5. A.A. Ardentov, A.P. Mashtakov, A.V. Podobryaev, Yu.L. Sachkov, *Symmetry method for optimal control problems on Lie groups*, in preparation.
6. A. Bellaiche, J. Risler, Eds., *Sub-Riemannian geometry*. Birkhäuser, Progress in Math., 1996, v. 144.
7. R. Bellman, *Dynamic Programming*, Princeton University Press, 1961.
8. R. Bellman, *Introduction to the mathematical theory of control processes*, vol. 1, 2. New York: Academic Press, 1967.
9. L.D. Berkovitz, *Optimal Control Theory*, Springer-Verlag, 1974.
10. G. Bliss, *Lectures on the Calculus of Variations*, The University of Chicago Press, Chicago and London, 1946.
11. A. Bloch, J. Baillieul, J. Marsden, *Nonholonomic mechanics and control*, Springer Series: Interdisciplinary Applied Mathematics, Vol. 24, 2003.
12. B. Bonnard, M. Chyba. *Singular trajectories and their role in control theory*, volume 40 of Mathématiques & Applications. Springer, Berlin, 2003.
13. U. Boscain, B. Piccoli, Optimal synthesis for control systems on 2-D manifolds. Springer SMAI, v.43, 2004.
14. U. Boscain, B. Piccoli, *A short introduction to optimal control*. In T. Sari, editor, Contrôle Non Linéaire et Applications, pp. 19–66. Hermann, Paris, 2005.
15. A. Bressan, B. Piccoli, *Introduction to the Mathematical Theory of Control*, American Institute of Mathematical Sciences, 2007.
16. R. W. Brockett, R. S. Millman, H. J. Sussmann, Eds., *Differential geometric control theory*. Birkhäuser Boston, 1983.

© The Author(s), under exclusive license to Springer Nature Switzerland AG 2022
Yu. Sachkov, *Introduction to Geometric Control*, Springer Optimization
and Its Applications 192, https://doi.org/10.1007/978-3-031-02070-4

17. F. Bullo, A.D. Lewis, *Geometric control of mechanical systems: modelling, analysis, and design for simple mechanical control systems*, Springer, 2005.
18. L. Capogna, D. Danielli, S.D. Pauls, and J.T. Tyson. *An introduction to the Heisenberg group and the sub-Riemannian isoperimetric problem*, Volume 259 of Progress in Mathematics. Birkhäuser Verlag, Basel, 2007.
19. A.A. Davydov, *Qualitative theory of control systems*. Translations of Mathematical Monographs. American Mathematical Society, 1994.
20. L. Euler, *Methodus inveniendi lineas curvas maximi minimive proprietate gaudentes, sive Solutio problematis isoperimitrici latissimo sensu accepti*, Lausanne, Geneva, 1744.
21. R.V. Gamkrelidze, *Principles of Optimal Control Theory*, Springer, 2013.
22. J.-P. Gauthier, I.A.K. Kupka, *Deterministic observation theory and applications*, Cambridge University Press, 2001.
23. A. Isidori, *Nonlinear control systems: an introduction*, Springer-Verlag, 1985.
24. B. Jakubczyk, W. Respondek, Eds., *Geometry of feedback and optimal control*. Marcel Dekker, 1998.
25. F. Jean. *Control of nonholonomic systems: from sub-Riemannian geometry to motion planning*. SpringerBriefs in Mathematics. Springer, Cham, 2014.
26. V. Jurdjevic, *Geometric Control Theory*, Cambridge University Press, 1997.
27. E.B. Lee, L. Markus, *Foundations of Optimal Control Theory*, New York: Wiley, 1967.
28. R. Montgomery, *A tour of subriemannian geometries, their geodesics and applications*, Amer. Math. Soc., 2002.
29. H. Nijmeijer, A. van der Schaft, *Nonlinear dynamical control systems*, Springer-Verlag, 1990.
30. L.S. Pontryagin, V. G. Boltyanskii, R. V. Gamkrelidze, E.F. Mishchenko, *Mathematical Theory of Optimal Processes*, New York/London. John Wiley & Sons, 1962.
31. L. Rifford. *Sub-Riemannian geometry and Optimal Transport*. SpringerBriefs in Mathematics, 2014.
32. Yu.L. Sachkov, Control Theory on Lie Groups, *Journal of Mathematical Sciences*, Vol. 156, No. 3, 2009, 381–439.
33. Yu.L. Sachkov, *Controllability and symmetries of invariant systems on Lie groups and homogeneous spaces* (in Russian), Moscow, Fizmatlit, 2007.
34. H. Schättler, U. Ledzewicz, *Geometric Optimal Control: Theory, Methods and Examples*, Springer-Verlag New York, 2012.
35. E.D. Sontag, *Mathematical control theory: deterministic finite dimensional systems*, Springer-Verlag, 1990.
36. H.J. Sussmann, Ed., *Nonlinear controllability and optimal control*. Marcel Dekker, 1990.
37. A.M. Vershik, V.Y. Gershkovich, Nonholonomic Dynamical Systems. Geometry of distributions and variational problems. (Russian) In: *Itogi Nauki i Tekhniki: Sovremennye Problemy Matematiki, Fundamental'nyje Napravleniya*, Vol. 16, VINITI, Moscow, 1987, 5–85. (English translation in: *Encyclopedia of Math. Sci.* **16**, Dynamical Systems 7, Springer Verlag.)
38. M.I. Zelikin, *Optimal control and calculus of variations*, Editorial URSS, Moscow, 2014 (in Russian) & 2010 (in Spanish).
39. M.I. Zelikin, V.F. Borisov, *Theory of chattering control with applications to astronautics, robotics, economics, and engineering*, Basel: Birkhäuser, 1994.
40. Yu. Sachkov, Left-invariant optimal control problems on Lie groups: classifications and problems integrable in elementary functions, *Russian mathematical surveys*, 77:1 (463) (2022), 109–176.

Some Important General Papers on Geometric Control Theory and Related Topics

41. A. Agrachev, Methods of control theory in nonholonomic geometry. *Proceedings ICM-94 in Zürich*. Birkhäuser, 1995, 1473–1483.
42. R. Brockett, System Theory on Group Manifolds and Coset Spaces, *SIAM J. Control Optim.*, 10 (1972), pp. 265–284.
43. R.W. Brockett, Control Theory and Singular Riemannian Geometry, In: *Hilton P.J., Young G.S. (eds) New Directions in Applied Mathematics*. Springer, New York, NY, 1982.
44. R.W. Brockett, Nonlinear Control Theory and Differential Geometry. *Proceedings ICM-82 in Warsaw*, Polish Scientific Publishers, 1984, 1357–1368.
45. W.L. Chow, Über Systeme von linearen partiellen Differential-gleichungen erster Ordnung, *Math. Ann.*, 117 (1939), pp. 98–105.
46. A.F. Filippov, On certain questions of the theory of optimal regulation (in Russian), *Vestnik Moskovskogo Universiteta*, 2: 1959, pp. 25–32.
47. F. Frobenius. Über das Pfaffsche Problem. *J. Reine Andew. Math.*, 82:230–315, 1877.
48. E. Hakavuori, E. Le Donne, Non-minimality of corners in subriemannian geometry, *Invent. Math.*, 206(3): 693–704, 2016.
49. R. Hermann. The differential geometry of foliations, ii. *J. Appl. Math. Mech.*, 11:303–315, 1962.
50. R. Hermann. On the Accessibility Problem in Control Theory. In: *International Symposium on Nonlinear Differential Equations and Nonlinear Mechanics*, pages 325–332. Academic Press, New York, 1963.
51. V. Jurdjevic and H. Sussmann, Control systems on Lie groups, *J. Diff. Equat.*, **12**, 313–329 (1972).
52. R.E. Kalman, Contributions to the Theory of Optimal Control, *Bol. Soc. Matem. Mexicana (Ser. 2)*, 5: 102–119.
53. A.J. Krener, A generalization of Chow's theorem and the bang-bang theorem to nonlinear control problems, *SIAM J. Control*, 12 (1974), 43–52.
54. R. Montgomery. Abnormal minimizers. *SIAM J. Control Optim.*, 32(6): 1605–1620, 1994.
55. T. Nagano. Linear differential systems with singularities and an application to transitive Lie algebras. *J. Math. Soc. Japan*, 18(4):398–404, 1966.
56. P.K. Rashevskii. Any two points of a totally nonholonomic space may be connected by an admissible line. *Uch. Zap. Ped. Inst. Liebknechta* (in Russian), 2: 83–94, 1938.
57. P. Stefan, Accessible sets, orbits, and foliations with singularities. *Proc. London Math. Soc.*, Ser. 3 29(4):699–713, 1974.
58. H.J. Sussmann, Orbits of families of vector fields and integrability of distributions, *Trans. Amer. Math. Soc.*, 180 (1973), 171–188.
59. H. Sussmann and V. Jurdjevic, Controllability of nonlinear systems, *J. Differential Equations*, 12 (1972), 95–116.

Papers in Which Particular Optimal Control Problems Are Studied

60. A. Agrachev, D. Barilari, U. Boscain, On the Hausdorff volume in sub-Riemannian geometry, *Calculus of variations and partial differential equations*, 43 (2012), 3–4, 355–388.
61. A. Agrachev, B. Bonnard, M. Chyba, and I. Kupka. Sub-Riemannian sphere in Martinet flat case. *ESAIM Control Optim. Calc. Var.*, 2: 377–448, 1997.

62. A. Ardentov, G. Bor, E. Le Donne, R. Montgomery, Yu. Sachkov, Bicycle paths, elasticae and sub-Riemannian geometry, *Nonlinearity*, 34 (2021) 4661–4683.

63. A. Ardentov, E. Hakavuori, Cut time in sub-Riemannian problem on Cartan group, *ESAIM: COCV*, 28 (2022) 12.

64. A.A. Ardentov, Yu.L. Sachkov, Maxwell Strata and Cut Locus in the Sub-Riemannian Problem on the Engel Group, *Regular and Chaotic Dynamics*, December 2017, Vol. 22, Issue 8, pp 909–936.

65. A.M. Arthurs and G. R. Walsh. On Hammersley's minimum problem for a rolling sphere. *Math. Proc. Cambridge Phil. Soc.*, 99(3): 529–534, 1986.

66. D. Barilari, U. Boscain, J.-P. Gauthier, On 2-step, corank 2 nilpotent sub-Riemannian metrics, *SIAM Journal on Control and Optimization*, 50 (2012), 1:559–582.

67. V.N. Berestovskii, I.A. Zubareva. Geodesics and shortest arcs of a special sub-Riemannian metric on the Lie group SO(3). *Sibirsk. Mat. Zh.*, 56(4): 762–774, 2015.

68. V.N. Berestovskii, I.A. Zubareva. Sub-Riemannian distance in the Lie groups SU(2) and SO(3). *Mat. Tr.*, 18(2): 3–21, 2015.

69. V.N. Berestovskii, I.A. Zubareva. Geodesics and shortest arcs of a special sub-Riemannian metric on the Lie group SL(2). *Sibirsk. Mat. Zh.*, 57(3): 527–542, 2016.

70. D. Bernoulli, 26th letter to L. Euler (October, 1742). In: *Correspondance mathématique et physique*, Vol. 2, St. Petersburg (1843).

71. J. Bernoulli, Véritable hypothèse de la résistance des solides, avec la demonstration de la corbure des corps qui font ressort. In: *Collected works*, Vol. 2, Geneva (1744).

72. I.Yu. Beschastnyi. On the optimal rolling of a sphere with twisting but without slipping. *Mat. Sb.*, 205(2): 3–38, 2014.

73. I.Yu. Beschastnyi, Yu. L. Sachkov, Geodesics in the sub-Riemannian problem on the group SO(3), *Sb. Math.*, 207:7 (2016), 915–941.

74. M. Born, *Stabilität der elastischen Linie in Ebene und Raum*. Preisschrift Und Dissertation, 1: 5–101, Göttingen 1906.

75. U. Boscain and F. Rossi. Invariant Carnot–Caratheodory metrics on S^3, SO(3), SL(2), and lens spaces. *SIAM J. Control Optim.*, 47(4): 1851–1878, 2008.

76. R. Brockett and L. Dai, Non-holonomic kinematics and the role of elliptic functions in constructive controllability. In: *Nonholonomic Motion Planning* (Z. Li and J. Canny, eds.), *Kluwer, Boston* (1993), pp. 1–21.

77. Y.A. Butt, A.I. Bhatti, Yu.L. Sachkov, Cut Locus and Optimal Synthesis in Sub-Riemannian Problem on the Lie Group SH(2), *Journal of Dynamical and Control Systems*, 23 (2017), 155–195.

78. L.E. Dubins, On curves of minimal length with a constraint on average curvature, and with prescribed initial and terminal positions and tangents, *American Journal of Mathematics*, vol. 79, no. 3, pp. 487–516, 1957.

79. B. Gaveau, Principe de moindre action, propagation de la chaleur et estimees sous elliptiques sur certains groupes nilpotent, *Acta Mathematica*, 139 (1977), pp. 95–153.

80. V. Jurdjevic, The geometry of the plate-ball problem, *Arch. Rat. Mech. Anal.*, 124 (1993), 305–328.

81. I. Moiseev, Yu. Sachkov, Maxwell strata in sub-Riemannian problem on the group of motions of a plane, *ESAIM: COCV*, 16 (2010), 380–399.

82. A.A. Markov, Some examples of the solutions of a special kind of problems in greatest and least quantities (in Russian), *Soobshch Kharkovsk. Mat. Obshch.*, 1887, v.1, 250–276.

83. O. Myasnichenko. Nilpotent (3, 6) sub-Riemannian problem, *Journal of Dynamical and Control Systems*, 8:4 (2002), pp. 573–597.

84. T. Pecsvaradi, Optimal horizontal guidance law for aircraft in the terminal area, *IEEE Trans. Autom. Contr.*, Vol. AC-17, No. 6, December 1972, 763–772.

85. A.V. Podobryaev, Yu.L. Sachkov, Cut locus of a left invariant Riemannian metric on SO(3) in the axisymmetric case, *Journal of Geometry and Physics*, 110 (2016) 436–453.

86. A.V. Podobryaev, Yu.L. Sachkov, Symmetric Riemannian Problem on the Group of Proper Isometries of Hyperbolic Plane, *J Dyn Control Syst*, (2018) 24: 391–423.

87. L. Saalschütz, *Der belastete Stab*. Leipzig (1880).
88. Yu.L. Sachkov, Symmetries of flat rank two distributions and sub-Riemannian structures, *Transactions of the AMS*, 2004, 2, 457–494.
89. Yu.L. Sachkov, Maxwell strata in Euler's elastic problem, *Journal of Dynamical and Control Systems*, Vol. 14 (2008), No. 2 (April), 169–234.
90. Yu.L. Sachkov, Symmetries and Maxwell strata in the ball-plate problem, *Sbornik Mathematics*, V. 201 (2010), N 7, 1029–1091.
91. Yu.L. Sachkov, Conjugate and cut time in the sub-Riemannian problem on the group of motions of a plane, *ESAIM: COCV*, 16 (2010), 1018–1039.
92. Yu.L. Sachkov, Cut locus and optimal synthesis in the sub-Riemannian problem on the group of motions of a plane, *ESAIM: COCV*, 17 (2011), 293–321.
93. Yu.L. Sachkov, Conjugate points in Euler's elastic problem, *Journal of Dynamical and Control Systems*, vol. 14 (2008), No. 3 (July), 409–439.
94. Yu.L. Sachkov, E.F. Sachkova, Exponential mapping in Euler's elastic problem, *Journal of Dynamical and Control Systems*, Vol. 20 (2014), No. 4, 443–464.
95. Yu.L. Sachkov, Maxwell strata and symmetries in the problem of optimal rolling of a sphere over a plane, *Sb. Math.*, 201:7 (2010), 1029–1051.
96. Yu. Sachkov, Conjugate time in sub-Riemannian problem on Cartan group, *Journal of Dynamical and Control Systems*, 2021, 27, 709–751.
97. Yu. Sachkov, A.Yu. Popov, Sub-Riemannian sphere on Engel group, *Doklady RAN. Mathematics, informatics, control processes.* 2021, Vol. 500, pp. 97–101.
98. H.J. Sussmann, G. Tang, Shortest paths for the Reeds–Shepp car: A worked out example of the use of geometric techniques in nonlinear optimal control. *Rutgers Center for Systems and Control Technical Report 91-10*, September 1991.

Other Cited Books and Papers

99. N.I. Akhiezer, *Elements of the Theory of Elliptic Functions*, AMS, 1990.
100. G. Citti and A. Sarti. A cortical based model of perceptual completion in the roto-translation space. *J. Math. Imaging Vis.*, 24(3):307–326, 2006.
101. J. Hadamard. Les surfaces a courbures opposees et leurs lignes géodésique. *J. Math. Pures Appl.*, 4: 27–73, 1898.
102. D.H. Hubel, T.N. Wiesel. Receptive fields, binocular interaction and functional architecture in the cat's visual cortex, *The Journal of Physiology*, 160:1 (1962), pp. 106–154.
103. S.G. Krantz, H.R. Parks, *The Implicit Function Theorem: History, Theory, and Applications*, Birkäuser, 2001.
104. D.F. Lawden, *Elliptic Functions and Applications*, Springer-Verlag New York, 1989.
105. R. Levien, The elastica: a mathematical history // Technical Report No. UCB/EECS-2008-103, 2008, P. 1–25.
106. A.E.H. Love, *A Treatise on the Mathematical Theory of Elasticity*, 4th ed., New York: Dover, 1927.
107. J. Petitot and Y. Tondut, Vers une neurogéométrie. Fibrations corticales, structures de contact et contours subjectifs modaux, *Math. Inform. Sci. Humaines*, 145 (1999), 5–101.
108. A.V. Sarychev, D.F.M. Torres, Lipschitzian regularity of minimizers for optimal control problems with control-affine dynamics, *Applied Mathematics and Optimization*, 41: 237–254 (2000).
109. S. Sternberg, *Lectures on differential geometry*, Prentice-Hall, Englewood Cliffs, N.J., 1964.
110. D.T. Suzuki, *Manual of Zen Buddhism*, 2015 Golden Elixir Press.
111. V.M. Tikhomirov, *Stories About Maxima and Minima*, Universities Press (India) Pvt. Limited, 1990.
112. S. Timoshenko, *History of Strength of Materials*, McGraw-Hill, New-York, 1953.

113. C. Truesdell, The Influence of Elasticity on Analysis: The Classic Heritage, *Bulletin American Math. Society*, 1983, v. 9, No. 3, 293–310.
114. F. Warner, *Foundations of Differentiable Manifolds and Lie Groups*, Springer, 1983.
115. E.T. Whittaker, G.N. Watson, *A Course of Modern Analysis*, Cambridge University Press, 1996.
116. S. Wolfram, *Mathematica: a system for doing mathematics by computer*, Addison-Wesley, Reading, MA 1991.

Index

© The Author(s), under exclusive license to Springer Nature Switzerland AG 2022

Yu. Sachkov, *Introduction to Geometric Control*, Springer Optimization
and Its Applications 192, https://doi.org/10.1007/978-3-031-02070-4

Printed in the United States
by Baker & Taylor Publisher Services